Superterrorism

Assassins, Mobsters, and Weapons of Mass Destruction

To Sarah and Bill
whose young minds
should not be gripped by terror
but, rather, allowed to run free
to explore the magic of the universe

Superterrorism

Assassins, Mobsters, and Weapons of Mass Destruction

Glenn E. Schweitzer

with

Carole C. Dorsch

Plenum Trade • New York and London

Library of Congress Cataloging-in-Publication Data

Schweitzer, Glenn E., 1930–
  Superterrorism : assassins, mobsters and weapons of mass
destruction / Glenn E. Schweitzer with Carole C. Dorsch.
    p.   cm.
  Includes bibliographical references and index.
  ISBN 0-306-45990-6
  1. Terrorism.   I. Dorsch, Carole C.   II. Title.
HV6431.S375 1998
303.6'25--dc21
                                                                98-28476
                                                                     CIP

ISBN 0-306-45990-6

© 1998 Glenn E. Schweitzer and Carole C. Dorsch
Plenum Press is a Division of Plenum Publishing Corporation
233 Spring Street, New York, N.Y. 10013

http://www.plenum.com

10 9 8 7 6 5 4 3 2 1

Printed in the United States of America

Contents

Acknowledgments

*T*he fingerprints of more than three hundred colleagues can be found throughout the pages of this book. I am a practitioner in promoting international cooperation, not a law enforcement official, narcotics agent, nor criminologist. Thus, I have relied heavily on the insights of specialists—in these fields and many others—in understanding the significance of the turmoil and violence I have encountered in many countries, indicators of major national security threats of the next century. Among these many mentors, I pay special tribute to the following (in chronological order from 1969 to the present):

- Senior American diplomats who guided my presentations at intergovernmental negotiations of nuclear arms limitations in New York, Brussels, and Geneva.
- Officials of our overseas foreign assistance missions (United States Agency for International Development) who provided both advice and armed guards to ensure I would avoid drug traffickers and insurgents in Latin America, Asia, and Africa.
- Fellow workers in the U.S. Environmental Protection Agency in Washington, D.C. and Las Vegas, Nevada, who joined me in developing the Toxic Substances Control Act and in responding to calls from fire and police departments around the country for assistance in coping with major accidents involving volatile and explosive chemicals.
- Faculty and students at Cornell University who repeatedly pointed out the devastating effects of governmental corruption and drug trafficking on economic development in Latin America and Asia.

- Scientists in the former Soviet Union and Eastern Europe who arranged for me to gain firsthand insights into many aspects of totalitarian rule, including participation in hushed seminars of oppressed dissidents in Moscow, tours of tightly-sealed East German institutes under the control of the Stasi security service, and meetings with rebel Romanian activists on the day before Nicolae Ceausescu was violently deposed and then shot.
- Colleagues from eight Western countries and dozens of Russian officials and scientists who supported my efforts to lead the establishment of the International Science and Technology Center in Moscow, which is devoted to converting weapons expertise to peaceful purposes.
- Coworkers at the National Research Council who participated in our assessments of the effectiveness of U.S. efforts to help contain nuclear and other dangerous technologies within the states of the former Soviet Union and in our establishment of new cooperative programs that are giving former Soviet biological weapons scientists careers devoted to public health concerns.
- Participants in terrorism studies of Georgetown University and George Washington University.

I am also grateful for the reviewer comments of Carol Rossi, Charles Fogelgren, Diane Schweitzer, Steven Michael Schwartz, Nina Ostrovitz, Mitchel Wallerstein, and Seth Carus.

And most important, I express my deepest appreciation to my wife Carole, who was my collaborator in writing every sentence of the book. She is a well-informed observer of today's international scene and a source of logic when it comes to structuring the discussion of complex issues. A skilled writer, she was tireless in her efforts to ensure that the text was fresh, understandable, and useful for our neighbors as well as for specialists in the field.

Prologue

The most ominous aspect of chemical and biological weapons is the movement of the front line of the battlefield from foreign soil to the American homeland.
Secretary of Defense William Cohen, 1997

Tomorrow's terrorist may be able to do more damage with a keyboard than with a bomb.
National Academy of Sciences, 1996

Saddam Hussein is like a punching bag: No matter how many times you hit him, he just keeps coming back in your face.
Time, 1997

If you live in or near New York City, Washington, D.C., Chicago, Illinois, Portland, Oregon, Tucson, Arizona, or even in smaller towns across America, such as Las Cruces, New Mexico, Norwalk, Connecticut, or Highland, Indiana, you have already become an unwitting victim of global terrorists. Their spheres of influence are spreading with the speed of advancing technology.

Perhaps alarm systems to guard homes and buildings, mobile phones to dial 911 or to stay in touch with loved ones will afford some degree of defense. But the steady intrusion of terrorism into everyday life raises profound questions that go far beyond the physical harm already suffered by a small number of random victims of terrorist violence—questions as to whether democracy and freedom can coexist with the dangerous activities they allow to operate unchecked.

You may rightly point out that it is not terrorists but rather thugs and common criminals who are provoking the pervasive security consciousness in our cities and towns. It's the carjackers and the robbers, not the terrorists, who have impelled owners of automobiles and homes in Louisiana to vote themselves the right to shoot criminals caught in the act. It's the neighborhood predators and dueling gangs, not the purveyors of political violence, that pose the threat to the youth of Chicago, Detroit, and New York.

Only when you negotiate the security gauntlet at any American airport or walk through the magnetic portal of a government building do you experience the kind of in-your-face adjustments in America's way of life directly attributable to terrorists. A delay of several minutes to answer a few routine questions from airline personnel and to place carry-on luggage on conveyor belts is no big deal, because such measures give the traveler some sense of security, however misplaced. Besides, you've already arrived early, as instructed by the airlines, to ensure that you do not miss your flight. And we all understand that special precautions are needed at federal buildings and at county courthouses, which occasionally attract disgruntled residents from our own communities and elsewhere.

At the same time, however, pulsing in the background of our daily lives is the backbeat of bulletins reporting real or imagined acts of terrorists. Such dramatic events and speculations foster feeding frenzies that regularly seize the headlines printed boldly across the tabloids, flashed on the screens of television sets, and even posted to the Internet. Terrorists, though not yet a major source of disruption in America, are pictured as quietly, but methodically, masterminding a mosaic of potential violence that will become a tapestry for life in the next century. In fact, many are well-organized criminals with deep pockets. Motivated by politics, religion, or just plain greed, they will not be easily dissuaded from perpetrating the most heinous acts.

During November 1997, a month selected at random, an informal survey looked at the frequency with which friends and family living in the previously named American communities were subjected to images and reports of terrorism. Each household was asked to tally the number of times during November that information about terrorism intruded into their everyday lives. From television news reports to the marquees of local movie houses to new book titles at libraries and bookstores covering everything from cybercrime to bomb threats, news of terrorist activity shadowed them like the most persistent stalker. On forty-eight separate occasions, friends in Las Cruces, New Mexico, were confronted with terrorist images. In Highland, Indiana, 181 hits were recorded. In New York City, the tally totalled ninety-nine, even though New York is a city where incidents of political and racial violence are so frequent they often go unreported. In Portland, Oregon, relatives, who are voracious readers, posted 272 items. And in the Washington suburb of Arlington, my own household, albeit sensitized to the topic, recorded 133 encounters indicating that more than four times daily we were reminded of the terrorist threat.

Of course, the results of the survey were highly dependent on the participants' interests, their amount of free time, their media indulgence, and reading habits. Also, it just happened that the survey coincided with the trials and legal troubles of Ramzi Yousef, who attempted to bomb airliners from the skies over the Pacific Ocean, Terry Nichols, who helped plan the Oklahoma City bombing, and Mir Aimil Kasi, the assassin of two CIA employees in Langley, Virginia. Also, the murder of tourists in Egypt and Saddam Hussein's face-off with United Nations (UN) inspectors punctuated the news. But watching these and still more terrorist dramas play out during the succeeding three months convinced me that November really did reflect the tenor of the times. We can conclude that even the most disinterested American finds it difficult to entirely elude the far reach of the terrorist threat.[1]

Thus, if jolting us out of our sense of well-being is the objective of terrorists, they are beginning to succeed—even though the actual bombing incidents, kidnappings, and other attacks in America have been few and far between. From a baseline of subliminal apprehension, some of us are becoming steadily convinced that fanatics, crusaders, and brilliant dissidents with distorted minds are at work in clandestine corners of the globe. Interestingly, we regularly twist distant threats into new forms of entertainment that channel our fears in perverse ways. We explicitly replicate terrorist havoc on our television screens and in local movie theaters. We even cheer for the success of the terrorists when their goal is to right the wrongs of evil governments.

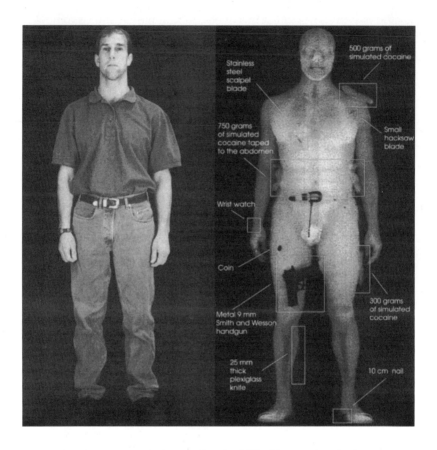

1. What do terrorists and smugglers look like? A scanning device exposes a harmless looking traveler's secret cache of drugs and weaponry.

Screen writers and novelists will not change course. They have seized a topic that mesmerizes all. Enough actual incidents confront us to instill plausibility into some of the wildest plots of all times.

America's first line of defense against the real acts of terrorism, which will undoubtedly multiply as the new millennium begins to unfold, is to understand the nature of the beast. To prevent the apocalyptic Hollywood scenarios from reaching reality, we need first a firmer understanding of the evolving characteristics of terrorism in its many dimensions. Only then can we take effective steps to protect ourselves, our institutions, and our society from becoming the victims so vividly portrayed in film clips that, at first glance, may mistakenly seem to be unimaginable. To do nothing, to retreat to cocoons of isolation, or on the other hand to overreact by moving toward a police state will admit defeat to terrorists who are bent on changing our democratic lifestyle, a lifestyle that in duplicitous ways they both despise and envy.

How serious is the current and future threat of terrorism? Recent reports of the Department of State indicate a steady decline in the number of worldwide incidents classified as terrorism, although incidents with large casualty tolls have become more commonplace. Also, one senior official from the Federal Bureau of Investigations (FBI) pointed out to me there were only two deaths from terrorist activity in the United States during 1996, while in contrast thousands of inner-city residents are murdered each year.[2]

However, most experts agree that these statistical indicators of terrorism incidents are transient at best and will likely be displaced by new statistics scored from another spate of events every bit as dramatic and deadly as the catastrophes in Oklahoma City, New York, or Tokyo. Indeed, I share the conventional wisdom with my Washington colleagues that the next big strikes are just over the horizon and that the surveillance systems of our law enforcement and intelligence agencies may not be capable of detecting them before they hit home. All those who argue that terrorism will remain little more than background noise (or convenient hype by government agencies seeking larger budgets to fend off a nonexistent enemy) simply are not in touch with developments in the Middle East, Latin America, and Asia, let alone the United States itself. As I intend to show, we need to recognize the lulls before each storm and realize they are only temporary. If we don't, we will never be fully prepared to either diffuse the next upheaval before it hits or respond when it actually does.

In this regard, the same FBI agent contended that even if there is a growing *latent* threat, we can rest easy since 2,600 FBI specialists are devoted full-time to keeping terrorists at bay. They are reinforced by many more specialists in other government agencies. But, with foreign terrorist organizations establishing outposts in the United States, with domestic dissidents becoming bolder in stockpiling armaments, and with HazMat (hazardous materials) response teams outfitted in space suits appearing in cities throughout the country, I am not convinced his confidence is completely warranted.[3]

Exemplifying the dangers, a well-respected American cell biologist boasts he has developed a simple, one-man scheme to impregnate the rugs of the White House with bacteria that spread bubonic plague—not only demonstrating that such a deed could be accomplished, but could be done rather simply and cheaply at that. He claims that all he—or any other well-trained biologist—needs is: (a) an airplane ticket to fly to an area in the United States where he can capture rodents infested with fleas that transmit the plague; (b) a few items from a local drugstore to stock the simple laboratory in his basement; (c) a surplus army gas mask to ensure he survives the process of extracting and packaging the deadly bacteria; and (d) a wheelchair that will gain him entry, along with genuinely disabled visitors, to the second floor of the White House. Conceivably the White House could soon become both an infirmary and a morgue, with rooms and offices under quarantine for several weeks. Not to worry about this scenario, he then adds. He claims he created this model to defuse it and is dedicating his life to countering, not perpetuating, such threats.[4]

A highly experienced professor of computer science offers several other frightening scenarios. He contends that cybersleuths are able to continuously recycle blitzes of bits and bytes to completely overwhelm the carrying capacity of important segments of the electronic networks controlling our transportation, communication, and electrical power systems. Also, designers of digital missiles are targeting financial institutions that may lack secure backup systems for their records; many institutions use the same computers for primary and secondary systems. Often their firewalls to protect records can be readily penetrated. The professor also cites risks in importing software from low-cost foreign labor markets that could be embedded with hidden programs designed to crash American systems on command from abroad. To emphasize the reality of cyberterrorism, he recalls the success of the Bulgarian "school" that instructed its students on how to plant electronic viruses in computers around the world in the late 1980s.[5] Their associates would then sell the software to combat the viruses.

Our counteroffensives are no less troubling. We may initially applaud technological advances that enable police to use electromagnetic cameras to spot weapons hidden under clothing of suspects at a distance of up to sixty feet.[6] However, this is only the beginning. Recognizing that plastic guns escape the airport magnetometers, electronic engineers are designing new types of scanning devices to see through clothing and reveal weapons caches. One technology is described as being so sensitive the security guard will be able to tell if a fully clothed man has been circumcised. Other technologies are being created to peer through thin walls of buildings and private homes—once considered our hallowed sanctuaries—in the search for contraband and suspected criminals.[7]

But we need not rely solely on conjecture in assessing the need for action to avoid future disasters. Actual events underscore terrorists' ability to fulfill their threats, while more and more fictitious plots simply embellish on past events or presage what can happen at any given moment.

The film *The Rock* promised widespread bodily immobilization resulting from bright yellow and green glass capsules of nerve agents being dropped on San Francisco. Audiences held their breath at the thought of an entire city being asphyxiated by debilitating chemical vapors. In the real world of Japan, an analogous event almost happened when the Aum Shinrikyo religious cult attacked a Tokyo subway station in 1995. Mistakes made by the cult limited the number of casualties to a dozen deaths and several thousand injuries. An error in their preparation of a batch of the nerve agent sarin reduced its potency. Moreover, their system for dispersing the gas was poorly designed. If not for these shortcomings, tens of thousands of passengers could easily have been killed in this busy subway system that carries over five million people each work day.[8] Later we learned that this cult had planned to export its sarin bombs to the United States as well. Had they not been stopped, who knows what the body count of Americans would have been on our own soil?[9]

The Towering Inferno and other fictitious precursors to the 1994 World Trade Center bombing showed in horrifying detail the effects that destructive forces can have on the high-rise symbols of America's cities. The devastation that Hollywood portrayed, however, was not as chilling as the anticipated results of the plot orchestrated by the World Trade Center bombers. Judge Kevin Duffy, who presided over the trials of the perpetrators, pointed out that not only had they attempted to topple the north tower into the south tower, like a pair of dominoes, but they also planned to unleash a poison gas attack. Had the sodium cyanide packed in with the explosives not burned as planned but vaporized instead, the cyanide gas would have been sucked into the north tower, killing all of the inhabitants, according to the judge.[10]

The potential damage from evil plots seems limitless. Holding a nation hostage to nuclear blackmail, hijacking the president's jet, undermining confidence in the banking system by destroying financial records, and disseminating diseases around the world on passenger airplanes are among the hottest threats making the Hollywood rounds. Unfortunately, some of yesterday's most far-fetched schemes could become reality within the next few years. Clearly, fact and fiction are entwined in a deadly embrace. In the absence of strong countermeasures, the consequences for society could be catastrophic.

Though modern society has become vulnerable as never before, coping with terrorism is hardly new. It is as old as our earliest civilizations, and fear over the increasing access to modern technologies by terrorists goes back many decades. The following could easily be a quotation from the most recent *Congressional Record*:

> Murderous organizations have increased in size and scope; they are more daring, they are served by more terrible weapons offered by modern science, and the world is nowadays threatened by new forces, which, if recklessly unchained, may some day wreak universal destruction.[11]

It is actually a warning, at the turn of the last century, of a British policeman who noted that the dastardly science of destruction had made rapid and alarming strides between 1858 and 1898.

Now, 100 years later, the tragedies around the globe are mounting: tourists in Egypt are shot, hacked, and mutilated; bus bombings in Israel make commuting to work as dangerous as playing Russian roulette; attacks on embassies cripple diplomatic efforts; and narcoterrorists force Third World governments to accept corruption as a way of life while at the same time destroying lives of children in America and other distant lands. Indeed, during the 1990s, terrorist acts have enjoyed an unwelcome familiarity in everyday life in far too many countries—Ireland, Israel, Bosnia, Sri Lanka, Pakistan, and Algeria, to name a few.

But, it is only the recent bombings in New York, Oklahoma City, and Atlanta, and the body bags returned to the United States from a demolished American apartment complex in Saudi Arabia that have shaken and shocked many Americans into an up-to-date vision of reality that includes such incomprehensible acts.

In short, the threat *is* very serious. Counteroffensives aimed at the likely perpetrators of attacks and preparations for immediate responses at the scenes of incidents are essential.

The number of terrorist groups around the globe plotting to settle what they consider to be unpaid debts of retribution with the United States and others has spiraled upward despite the decline in incidents worthy of reporting. Approximately two thousand terrorist groups, small and large, threaten the world today; probably 800–1,000 of them are very active, sharing many of the same deadly goals.[12]

As these groups proliferate and strengthen, their requirements for weaponry, communications networks, and concealment techniques also increase. With rising operating costs, some are beginning to collaborate with organized crime. These crime syndicates currently spin more than $500 billion per year of laundered money through the financial turnstiles of the world, primarily for the benefit of themselves but increasingly as rewards for their terrorist collaborators as well. The line separating international terrorism and organized crime is, in fact, fading as quickly as the color of the well-washed currency. And the global nature of American interests pulls the United States right into the vortex of the danger.

Ironically, terrorist groups and their agents are establishing bases in the United States where freedom of movement is a hallmark of our democratic system. The safeguards against the entry of such dangerous new residents are unreliable due to bureaucratic indifference or to policies that try to respond to contradictory agendas. For example, in the rush to naturalize immigrants, presumably so they could vote in the 1996 presidential election, the Immigration and Naturalization Service waived FBI screening procedures for 150,000 applicants for American citizenship.[13] Did even one or two terrorists with designs on American targets slide in while no one was looking? Meanwhile, as was true during the crisis over UN access to facilities in Iraq at the end of 1997, Middle East diplomats have been accused of facilitating the activities in the United States of representatives of terrorist groups allied with their governments.[14]

We must not forget our own homegrown terrorists. Past incidents of brutal behavior have usually resulted from single-issue extremists using violent means to advance their agendas, ranging from prohibiting abortion to defending animal rights. More recently, responses to the disastrous events at Waco and Ruby Ridge have pumped up the membership rolls of state militia groups and spawned as many as twenty thousand sympathizers in angry fringe groups ready to promote equally threatening scenarios.[15] While not all militias may have terrorism in mind, the apparent connection between militia goals and the Oklahoma City bombing sends a warning about the capabilities of our own citizens.

Indeed, our Achilles heel may lie here at home. Beyond the spread of militias, we must consider the possibility of more radical groups in the United States—with immediate access to the best and worst of American technologies—becoming unwitting allies of well-financed foreign militants whom they may even resent but who target the same U.S. institutions for different reasons.[16]

Finally, as our computer science professor underscored, cyberterrorism, the newest genre in the technoterrorist library, is stirring anxieties. The thousands of attempts by various, motivated computer hackers to penetrate the inner sanctums of the Pentagon, repeated efforts to manipulate the records of America's largest banks, and plots to confuse the air traffic control system have greatly perturbed even the most hardened law enforcement officials. Will this new form of information warfare, whether instigated by terrorists or others, provide the next great theater of conflict?[17]

Many university libraries house hundreds of well-documented books, journal articles, and research papers on terrorism. Some recent publications provide a broad philosophical and historical setting on the topic; some are directed to very specific threats; and some chronicle solved and unsolved incidents. Most are intended for professionals in the field.

The accounts of current incidents, particularly those touted almost daily by both well-respected newspapers and lesser-regarded tabloids, provide riveting sound bites that often short-change the seriousness and complexities of the issues. The ferocity of decades of hatred that crystallizes into kidnappings and bombings cannot be captured simply in headlines, nor can the massive threat from the release of contagious germs be portrayed in a single photograph of an infected victim.

My purpose is to provide an easy read for the general public that covers many of the most critical aspects of this ever-more-complicated topic. My emphasis is on the international dimension of the danger to the United States, a dimension frequently masked by sometimes unnecessary security classifications and confused by spotty reporting. This overview is liberally punctuated with firsthand accounts of incidents around the world that help document the scope of the problem.

Much of the discussion is directed to immediate threats and responses. But a second theme is no less important. We need a long-term strategy to counter terrorism in its many new forms. As we will see, the battle will last for decades, and indeed longer, with a U.S. commitment on a level that rivals our expenditures to ensure we did not lose the Cold War. This strategy must maximize the likelihood that terrorists will be deterred, and when necessary, apprehended in an increasingly borderless world. It must also provide alternative employment and lifestyle options for disenfranchised people with legitimate needs. We cannot accommodate those who simply seek power for the sake of power.

A very brief review of the origins of modern terrorism and of developments that have driven its growth open the discussion. I describe some of the new tools of the terrorist trade and show the magnitude of the dangers. I explore how criminals are directing international flows of weapons, know-how, money, and dangerous materials across many common intersections, with opportunities for illicit acts to reinforce one another. In short, the synergy of the terrorism and organized crime partnership has the potential to become far more than the sum of its parts. It also evokes the danger of a new level of violence—superterrorism. While an accepted definition is still evolving, for the purposes of this book superterrorism will include acts based on advanced technologies that cause massive destruction or high death tolls. This means attacks involving chemical or biological agents, other than minor poisoning incidents, or nuclear material configured as a weapon or packed around explosives. Also included are incidents based on plastic explosives and cyber attacks targeted at electronic networks that service our security, economic, and emergency support systems.

I highlight important conclusions set forth by others, while distilling from recent history and from evolving theories of terrorism a few lessons learned that will likely affect the American way of life in the years ahead. We all hope that none of our neighborhoods will have to contend with massive devastation; however, lesser forms of destruction, though not terrorism per se, have already crept into our cities, sometimes making an unexpected turn onto the "wrong" street an invitation for disaster.

Of course, I cover the demise of the Soviet Union whose decline has been of special concern to the world at large. Regrettably, its nuclear arsenal attracts far greater interest today in the West than its struggle to introduce newfound democracy and economic reforms. How we respond to the unfolding drama of leakage of nuclear and other dangerous technologies from the territory of the former Soviet Union is vital to the global community.

War and terrorism, urban crime and terrorism, and financial greed and terrorism are pairs of activities that often live in close proximity. Thus, I stress the need to combat many forces of evil, as both independent and consolidated threats. Just as the purveyors of weapons and materials of mass destruction, instigators of political violence, barons of organized crime, drug traffickers and associated money launderers, and corrupt government officials can reinforce one another, the measures for defeating them all must be equally complex. Unfortunately, on occasion a few employees of long-trusted U.S. agencies have also participated in illegal activities that have benefited America's adversaries, making the job of ethical, dependable employees even more difficult.[18]

I present a few suggestions for intensifying the struggle against new types of threats. As the terrorists become more clever and creative in their zeal to generate fear, we need to become equally imaginative and adept at thwarting their schemes. For the long run, we must begin *now* to address in a far more serious way than we have done previously the environments and issues that serve to fuel terrorism. Some experts deplore what they consider to be naivete in those of us who advocate attacking what has been referred to as root causes of criminal and terrorist activity. They argue that early theories about root causes have been discredited and that terrorists must simply be treated as criminals. In my view, however, there are large groups of people who have legitimate needs that when met would no longer be motivated to participate in or support crime and terror. These are the targets of the programs and initiatives I recommend and they must be dealt with lest our legacy to future generations be but a lifetime of fear, confrontation, and combat—and in the worst case, a lifetime abruptly terminated by a nuclear holocaust or a worldwide plague.

Still, we must realize that much of the hold terrorists have over us is a psychological one. Any level of victory over terrorism lies in understanding its latest incarnation as best we can, defending ourselves against it as well as we can, and—in the end—living with the fact that any strategy will fall short of providing total confidence that we are safe.

As Lenin stated (and ruthless leaders from Robespierre to Qaddafi have demonstrated), the goal of terrorism is to terrorize. If we refuse to capitulate to that end, we have won no small victory. Knowledge, vigilance, and well-conceived defense strategies will go a long way in denying victory to the terrorists of the world. Meanwhile, we must find new ways to redirect terrorists into more constructive occupations.

CHAPTER 1

Modern Mutations of Terrorism

Our earth is degenerate in these latter days. Bribery and corruption are common. And the end of the world is evidently approaching.
Assyrian tablet, 2700 B.C.

In 1605 a terrorist tried to blow up the British House of Commons. He used twenty-nine barrels of gunpowder. Today the job could be done with a small plastic charge concealed in a briefcase.
Department of State briefing, 1997

However democratic a society, however near to perfection the social institutions, there will always be alienated and disaffected people claiming that the present state of affairs is intolerable, and there will be aggressive people more interested in violence than in freedom and tolerance.
Walter Laqueur, Georgetown University, 1987

The hooded figure waits in the shadows, totally still except for his nervous tapping of his foot. His eyes dart from right to left, sweat can be seen beading on his face—or the little of it exposed to the moonlight. Suddenly his victim emerges from the doorway to his left, as he knew he would. He has studied the man's routine night after night; he knows the exact moment to strike. Out of the shadows he hurls himself at the unsuspecting victim, plunging a knife into the man's chest. He disentangles the brightly stained dagger from the bloody clothing of the groaning man who slumps to the ground. Looking in both directions and readjusting his hood, the figure races off into the darkness.

Is this a scene from the outskirts of Berlin where a Middle East operative has evened the score with an informant who revealed the complicity of the Iranian government in terrorist bombings? Perhaps an Algerian rebel in Paris has sent yet another warning signal to the French authorities over their support of the ruling regime in his North African homeland. Or maybe a Russian hit man was hired to eliminate a Moscow export control official blocking a Colombian cartel bazookas-for-cocaine trade. Any of these explanations could easily fit into the script, with the result of the age-old terrorist tactic always being the same: a life snuffed out by a quick stab for an unstated cause.

Whether retribution or intimidation, such clandestine and premeditated acts of murder have commonly dotted our world's history. Political separatists and religious extremists, often relying on bands of marauders and hired killers, have terrorized humanity since the earliest civilizations. Rogue states, oppressive governments, and organized crime groups have also conspired to callously eliminate opponents who dared block their roads to wealth and power. To trace the trends of such brutality and injustice is to revisit the entire story of humankind. Unfortunately, all signs point to an eternal life for this most dastardly form of employment—the political murder-for-hire.

Several main events from history provide some perspective on recent permutations and on the responses of societies in addressing, and sometimes successfully containing, terrorism. Critical benchmarks along the way, ushered in by new technologies that upgraded weapons from knives and firebombs to guns and dynamite to nuclear devices and engineered disease viruses, have signaled a rise in the stakes of terrorism.[1]

The U.S. State Department's definition of terrorism was adopted in the wake of the 1972 tragedy at the Munich Olympics where Palestinian terrorists from the Black September Organization brutally murdered nine Israeli athletes:

> Terrorism is premeditated, politically motivated violence perpetrated against noncombatant targets by subnational or clandestine agents, usually intended to influence an audience.[2]

Held up to the mirror of history, this definition describes many actions of disaffected populations who, over the millennia, have continued to seek retribution. It leaves out a large number of other acts of political terror—in particular, heinous acts perpetrated openly by governments (or government-sponsored organizations) and attacks aimed at government forces.

Premeditated violence directed against both armies and general populations characterized the actions of rebellious forces, seeking new political orders, that destroyed the early civilizations of the Dark Ages, triggered the decline of the Egyptian Old Kingdom, brought down the states of Greece and Crete, and toppled the Roman Empire.[3]

Historians point to a common factor present in such great catastrophes perpetrated by armies or terrorists, as definitions dictate. Civilizations were defeated when the spread of metals technology enabled the forces of barbarism to arm themselves to such a level that they could successfully wreak havoc on both the ruling elites and the common people. In the last resort, civilizations stood or fell, not by covenants, but by the sword.[4] The impacts of technology that destroyed these civilizations provide an interesting parallel with our recent anxieties over the possibility of devastating nuclear incidents.

Between the eleventh and thirteenth centuries, a terrorist group of fanatical Arabic tribesmen lived in an area ranging from Persia to Syria. Called the Hashashinim, they gained notoriety for their brutal deeds conducted while they were high on hashish. *Assassin,* an Anglicized version of their name, has stood for political murder ever since—and with its obvious political intent, fits the State Department's definition. And, yes, terrorism and drugs were linked long before their twentieth century nexus in Latin America.[5] The Assassins, whose weapon of choice was the dagger, were willing to die for their cause and often committed murders in front of witnesses, ensuring their own immediate capture and conviction. They viewed the killings as a holy mission—much like ultra-extreme Islamic fundamentalists of today who believe their deeds will be rewarded in the afterlife.[6]

A father of modern terrorism was Maximilien Robespierre, who set up the Committee of Public Safety that ruled France during the turbulent years following the French Revolution. This dictatorship was a forerunner of the twentieth-century governments that have erected state-sponsored oppression in the name of codes of conduct: Stalin's Russia, Hitler's Germany, and Pol Pot's Cambodia. Robespierre unleashed his Reign of Terror between 1793 and 1794, justifying it as the only way to save the revolution from anarchy at home and from invasion by other European monarchs.[7]

The debut of Robespierre's guillotine signaled a new role for technological innovation in terrorist weaponry. It beheaded more than seventeen thousand people, while an additional twenty-five thousand political enemies were killed by other methods.[8] The broader goal of this brand of terrorism and its bloody public executions was clearly to create "nervousness, apprehension, and fear of secret denunciation that haunted thousands whom the guillotine did not touch."[9]

Throughout history, the demarcation between terrorism and what Robespierre might label as necessary violence and others would classify as genocide has usually been determined through the eyes of the beholder. Despite the State Department's well-worn definition, appropriate descriptors of terrorism and terrorists remain elusive. One source cites 212 definitions worldwide, with ninety in use by governments and well-established institutions.[10] While legal systems often require rigidity in such definitions in mandating punishments, governments sometimes settle for considerable elasticity. Such elasticity may be the reason why official definitions remain in place indefinitely. But as we will see, new types of terrorist threats now demand broad agreement on at least a minimum threshold for classifying acts as terrorism with appropriate punishments. And because of new partnerships between criminals and terrorists, such a threshold must encompass certain criminal behavior previously relegated to other less stringent types of legal redress.

Another benchmark in the evolution of terrorism gained the attention of our founding fathers. The pirates of the Barbary Coast states (Morocco, Algiers, Tunis, and Tripoli) inked their mark on the terrorism timeline by demonstrating that trading hostages for large ransoms was good business.[11] In 1795, a significant arms-for-hostages deal was struck between pirates and America. It included a cash payment, annual tributes, and naval armaments and frigates—a benefits package that eventually totaled nearly $1 million for Algiers. Though a few hundred American hostages were released unharmed, the Barbary Coast threat loomed large for several decades.[12]

In truth, the United States was born of an atmosphere of political violence—although we hate to tie such onerous terminology as terrorism to a heritage that Americans proudly embrace. I am sure that historians prefer to describe early atrocities as armed insurrections, military actions, or savage warfare, but not as terrorism. Nevertheless whether breaking from the British, conquering the Indians, or suppressing the slaves, the founders of the United States often resorted to strategies and tactics that by many of the 212 definitions could certainly be described as terrorism based on political intentions. Once the union was in place and the Civil War had faded into history, it was easier to distinguish between the legitimate power brokers and the renegades resorting to violent tactics.

In the late nineteenth century, discrimination and ethnic rivalry spawned a new brand of terrorism to be practiced by adversaries in labor–management conflicts. An example of the times was the confrontation between the Philadelphia and Reading Coal and Iron Company and the Molly Maguires, a group of Irish coal miners who reacted violently to what they justifiably considered discrimination and exploitation. They vented their wrath on not only the owners of the mines, but on fellow workers of Welsh and German extraction as well. The Haymarket Square bombing in 1886 was the first of several bloody altercations between factory police and militant miners and steelworkers. While the role of the Molly Maguires in such disturbances may have been exaggerated by the management of the mining companies for their own purposes, the linkage of violence to a particular ethnic group at that time was clear.[13]

Labor rumblings also were linked to the killing of Governor Frank Steunenberg of Idaho in 1905, presumably by William Haywood. One of the founders of the often-violent Independent Workers of the World, Haywood was acquitted of the crime for lack of evidence. Similarly, a lengthy period of union-management disputes and strikes at the *Los Angeles Times* was followed by the blowing up of the *Times* building in 1910.

Once again, technological innovation boosted the violence quotient of militants. Alfred Nobel's invention of dynamite in

1867, while intended for peaceful use, gave labor agitators a new weapon to use against their perceived opponents: big business and government.

During these times of conflict and confrontation in the United States, violence was limited in scope, and there was no movement to overthrow the government or to change the country's political system. The labor dissidents and others had smaller fish to fry and would work within the overarching system.[14]

Meanwhile, as the nineteenth century drew to a close, hot spots bubbled up around the world. It seemed no august person was safe from the attacks of individual anarchists bent on stirring up chaos. In 1894, an Italian anarchist assassinated French President Sadi Carnot. In 1897, assassins fatally stabbed Empress Elizabeth of Austria and killed Antonio Canovas, the Spanish prime minister. In 1900, the King of Italy, Umberto I, was felled in an anarchist attack. During the following year an American anarchist killed U.S. President William McKinley. One historian recently observed:

> If in the year 1900 the leaders of the main industrial powers had assembled, most of them would have insisted on giving terrorism top priority on their agenda, as President Clinton did at the Group of Seven meeting after the June 1996 bombing of the U.S. military compound in Dhahran, Saudi Arabia.[15]

It can be argued that some, if not all, of the assassinations noted above were carried out by individuals rather than by groups trying to impose new political agendas. Therefore, a few arrests could have solved the problems. However, while anarchists may have preferred to act alone, they were part of a movement of the times. Understandably, the public was both fearful of and fascinated by the secret and mysterious character of those who would commit such acts, sensing that such actions could occur again.

Much of the terrorism in the mid-twentieth century took place in areas remote from the mainstream of American consciousness. It was concentrated in conflicts in the Middle East, India, Ireland, and other distant lands. Although Middle East terrorism preceded the founding of the state of Israel by several decades, that development immediately precipitated a war over the very existence of the new state. This battle directly affected many Americans who had strongly supported the establishment of Israel and its right to peace. Dueling factions in the region remain in a deadly face-off to this day. In recent years, with the advent of CNN and its riveting footage of bloody confrontations around the world, Americans have had very good seats from which to observe the carnage from Jerusalem to Belfast to the depths of Africa.

For Americans, it took another technological advance—international air travel—to usher terrorist activity into our own backyards. The popularity of air travel quickly spawned a spate of hijackings to bring attention to conflicts elsewhere, be it Cuba, other regions of the Middle East, or Eastern Europe. Dealing with such ordeals, particularly when they involved the safety of American planes and passengers, filled the calendars of presidents from Kennedy to Reagan.

The year 1968 is often acknowledged as the beginning of a new era of international terrorism as skyjacking incidents swiftly riveted the attention of Americans to various parts of the globe. The diversion to Algeria of an El Al flight by the Popular Front for the Liberation of Palestine was a major event in the new programming. While the media cast ultra-extremist Muslims as dangerous bandits, in the Muslim world Algeria became a beacon of independence as it strove to break away from colonial rulers.

Additional dramatic hijackings seized the spotlight by providing mesmerizing visuals, such as the nabbing by Palestinian extremists of four separate aircraft bound for New York in September 1970. The subsequent blowing up of several of the absconded jumbo jets at Dawson's Field in Jordan attracted attention around the world—and fanned emotional flames of both the haves and have-nots.

The murders at the Munich Olympics had a lasting effect on Americans. It was truly a wake-up call for the U.S. government to begin shaping more aggressive counterterrorism policies, for implementation here and abroad. In October 1972, the Nixon administration created an official organizational structure to deal with the terrorist threat, the Cabinet Committee to Combat Terrorism—a structure that has since changed in name and charter, but not in stature as it remains an important focal point for governmental action.[16]

The 1970s ended painfully for Americans with several dozen American diplomats held hostage in Teheran. Televised snippets of their captivity kept audiences on edge across the United States until they were finally released at the beginning of the Reagan administration.

In the 1980s terrorism reached yet another level described as the era of conclusive events—bombings and other types of killing that happened so quickly that they obviated effective responses. Plastic explosives, in particular, became the powerful new weapons of choice—easy to conceal, yet of great lethality.[17] Another spectacular event, however, demonstrated that inconclusive events were still with us. Palestinian terrorists seized the cruise ship *Achille Lauro* off Port Said, Egypt, in October 1985. They held 413 hostages for two days. Then, after tossing a handicapped American to his death in the sea, they surrendered to Egyptian authorities.[18]

Throughout these years, most trouble continued to erupt elsewhere on the planet. In 1993, the terrorist drama played out in the bombing at the World Trade Center in New York City followed in 1995 by destruction of the Murrah Federal Building in Oklahoma City. Third-party terrorism had reached our own shores, with Middle East terrorists targeting the United States for its support of Israel and then home-grown terrorists expressing their own political notions.

The shift in centers of violence demonstrated the gradual movement of terrorism from national to global platforms during this century. Today, in an easily navigable world, the distinction between domestic and international terrorism is a tough call for even the most discerning referee. The worldwide availability of information about tactics and technologies, the expansion of international criminal support systems, and the growth of illicit sources of supplies and money have brought different types of criminals with common interests closer together, creating a melting pot of terrorists determined to use modern weaponry for hostile purposes.

Another international dimension of terrorism victimizes American citizens living or traveling overseas. In 1996, 109 terrorist incidents worldwide involved U.S. citizens and their interests. Ninety-two acts specifically targeted Americans.[19]

Americans should quite correctly feel more vulnerable to attacks from militant groups within or outside the United States today than did our ancestors 100 years ago. Nonetheless, in both timeframes the impact of terrorism on the general public has been more psychological than physical as the actual number of victims on record is small. Of course, that is not to say that every life doesn't count. At any rate, experts warn of the growing potential of terrorism with a message of ominous portent:

> The arms available to terrorists, the skills with which they use them, and, not least, the organizational techniques with which these weapons and skills are deployed are all improving at a fast and accelerating rate—a rate much faster than the countermeasures available to civilized society.[20]

In a sense, terrorism is gradually becoming an undeclared but open war with a host of perpetrators who can only be contained by high-energy international alliances of the world community. As lawlessness becomes global, laws can no longer be tightly constrained by borders when pursuing terrorists, law enforcement agencies must not only "get their asses," but must "get their assets," which often are shielded in safe havens.[21]

As we try to extract from recent history an accurate description of terrorism, our assumptions are constantly undermined. Many researchers and professionals have tried to capture the essence of this elusive subject, often with frustrating and inconsistent results.

In 1987, the internationally acclaimed scholar Walter Laqueur of Georgetown University set forth what he considered to be distortions in common beliefs about the character of terrorism. In other words, he felt that much could be gained by defining what terrorism is not. For example, he challenged theories that terrorism increases in response to the growth of repression, that terrorists are fanatics driven to despair by intolerable conditions, and that terrorism can happen anywhere. He argued that repression often quashes terrorist capabilities and ambitions, that terrorists are quite rational, and that certain areas are far more prone to terrorist attacks than others. Also, he doubted contentions that state-sponsored terrorism was becoming far more dangerous than

previous terrorist movements, and he questioned whether grievances, stresses, and frustrations underlying terrorism could ever
be put right.[22]

Then he drew an important conclusion:

> Statistics do not bear out the contention that terrorism is steadily
> rising at an alarming rate, but there are other weighty argu-
> ments concerning the peril of terrorism. The number of victims
> may be small, but terrorism is designed to undermine govern-
> ment authority; and it may have this effect by showing that
> democratic governments are unable to respond effectively.[23]

In contrast, most political scientists have chosen to explain
terrorism and terrorists by describing their characteristics. For
example:

- Terrorism provokes a fear and insecurity deeper than any other
  form of violence, striking innocent victims randomly and without warning; we feel defenseless against such attacks.
- Terrorists attempt to discredit governments by demonstrating
  their inability to protect their citizens.
- Terrorists use violence in an increasingly scattered way to express protest and rage, to advance fanatical religious agendas,
  and for other pathological reasons.
- Even though terrorism has almost always failed as a strategic
  weapon, it has cost the American government billions of dollars in direct costs, has damaged American economic interests,
  and has harmed allies and friends around the world.[24]

As to the weapons preferred by terrorists, in the early 1990s a
consensus seemed to emerge among experts that most criminals
will rely, as they have for more than a century, on the gun and the
conventional bomb. They prefer the "sophistication of simplicity"
and the availability of easily disguised weapons arsenals. We see
support for this conclusion as Algerian rebels continue to make
grenades from soda cans filled with gunpowder and wrapped in
tape. The growing legions of "amateur" terrorists, however, will
experiment with a variety of more deadly weapons, according to
conventional wisdom. Professional terrorists are rapidly gaining
the wherewithal to use the most advanced technologies when
they have the backing of rogue states. Thus, future terrorists will

have little difficulty crossing the threshold of high-tech weaponry, including acquiring weapons of mass destruction.[25]

The foregoing perspectives address many of the predominant concepts that have provided the backdrop for counterterrorism debates during the 1990s. The validity of some of the contentions may be questionable, and their significance will undoubtedly change as this millennium draws to a close. But because there are so many different types of terrorists and terrorism, all of the concepts—including ones criticized by Lacqueur as being misguided—will have some relevance in the years ahead.

Curiously missing from many discussions of the motivations propelling terrorists is the power of their most recent prop—money—either to finance expanded operations or to satisfy personal greed. Ideological, religious, and ethnic concerns will drive much of their activity. But the quest for money is equally significant as an aim of terrorist acts, sometimes as an intermediate step to finance ambitious plots for ideological, religious, or ethnic retribution.[26] (See Figure 2.)

In the past, terrorists might well have conducted their missions on the cheap with little concern over profits. Now, with the increasing cost of technology and the growing frequency of payrolls entwined with terrorist activities, the importance of money has become so pervasive that terrorist groups whose activities are limited exclusively to making political statements are becoming scarce. Meanwhile, new alliances of terrorists and organized criminals have alerted both groups to opportunities for marrying theft and intimidation, with the goal of cashing in on the increasingly exposed financial assets of prosperous American, European, and Asian institutions. The soft underbellies of such institutions, wrapped in electronic circuitry, are particularly inviting targets of vulnerability. While terrorism may still be the poor man's form of warfare for some groups, well-financed organizations with global connections will also be major players on the terrorist landscape of the future.

The fascination of experts and the general public with the psychological makeup of those who commit dastardly acts has led to many studies and much speculation as to the characteristics that set terrorists apart. On the world's Most Wanted List there is an endless cast of characters as scary as any of the grotesque suspects invented by novelists and screen writers.

Characterizations of such violent people include the following:

> The *professional terrorist* has a deep-seated resentment of all or certain forms of government and society and dedicates himself to their destruction. His profound hatred of society is often rooted in childhood experiences such as demanding, authoritarian parents who may indulge in political indoctrination and violence and who may also set perfectionist standards which the child cannot possibly attain. He likes to participate in carefully planned violence and even be present at the times of his crimes. Professionals have no political affiliations and simply work for money.
>
> The *political terrorist,* on the other hand, seeks to reform society. He has a love–hate relationship with society which leads to frustration, rage, and mental conflict. It is his duty to wreak vengeance on those he holds responsible for the evil political system of the country in which he lives. He is prepared to use violence to attain his political objectives but does not like to witness its results or meet his enemies face to face. His frustrations drive him to kill those whom he considers to be guilty or innocent with the object of compelling the authorities to give in to his demands.[27] (This description deserves a place in the biography of convicted Unabomber Theodore Kaczynski.)

Other experts add that those who do not shrink from assassination and bomb-throwing tend to be alienated young adults, with a crusading sense of mission. They come from middle-class or upwardly mobile respectable working-class families that may have strong religious roots. Terrorism is an equal-opportunity employer as evidenced by many females who have been terrorists.[28]

However, given the many types of terrorism and the hundreds and perhaps thousands of groups of terrorists, such broad generalizations cannot possibly embrace all motivations. Interviews of terrorists in Europe provide more specific insights as to the characteristics of members of at least a few organizations.

British researchers determined during their interviews with terrorists from the Provisional Irish Republican Army, the Red Army Faction in Germany, and the Red Brigades in Italy that they showed few if any of the attributes of clinical abnormality and no pathological qualities. They are essentially unremarkable people with no discernible personality consistencies, and they are generally undistinguished in appearance and manner. Also, the researchers noted the importance of an organizational environment providing paternal-type support for their activities.[29]

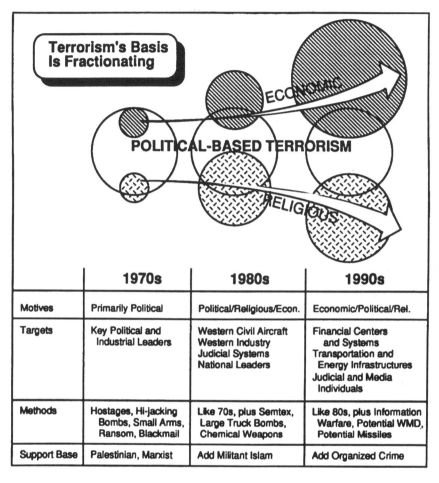

	1970s	1980s	1990s
Motives	Primarily Political	Political/Religious/Econ.	Economic/Political/Rel.
Targets	Key Political and Industrial Leaders	Western Civil Aircraft Western Industry Judicial Systems National Leaders	Financial Centers and Systems Transportation and Energy Infrastructures Judicial and Media Individuals
Methods	Hostages, Hi-jacking Bombs, Small Arms, Ransom, Blackmail	Like 70s, plus Semtex, Large Truck Bombs, Chemical Weapons	Like 80s, plus Information Warfare, Potential WMD, Potential Missiles
Support Base	Palestinian, Marxist	Add Militant Islam	Add Organized Crime

2. Growing Economic Motivations for Terrorism

The most disturbing aspect of these interviews was that the terrorists neither questioned nor renounced violent acts. As one respondent, sounding like other types of hardened criminals, flatly stated:

The only way to avoid the trauma of killing is to go on doing it. After a time you don't really think too much about it.[30]

An alternative take on terrorism is:

Once engaged, the terrorist seems to cross a number of boundaries. Political purity becomes compromised in the reality of terrorist life, with its mistakes, bad fortune, or simple inefficiency which results in the wrong victims. Such things as excitement or social approval carry the individual along, confirming and sustaining his or her involvement.[31]

These and other case investigations bring out different aspects of terrorism. Collectively they confirm the long-standing conclusion of most experts that while sharing some common attributes, each terrorist group is different and must be studied in the context of its own national culture and history.[32]

Terrorists acting under instructions from the government of a rogue state might have a makeup entirely different from members of truly independent groups. While these government agents could be characterized as professional terrorists, they might also be better likened to soldiers carrying out orders from their commanders. It would seem unrealistic to lump military conscripts together with volunteer terrorists.

The real fear is that the latest terrorist mutants are whetting their appetites with the power of weapons of mass destruction or with the instantaneous global reach of the computer. The danger lies not only in the devices themselves but in the minds of such technology-oriented mercenaries who, perhaps like proverbial mad scientists, will be driven to witness large-scale results of their latest experiments.

An up-to-date view by one terrorist of his mission of destruction was provided by the World Trade Center bomber Ramzi Ahmed Yousef during his sentencing to life imprisonment in January 1998. It is not certain whether he truly believed his bitter accusations or whether he simply wanted to justify his heinous behavior, but he was very clear in proclaiming he was proud to be a terrorist with America as his target. As expected, he condemned Israel. He then accused the United States as also being an inventor of terrorism, citing the atomic bombing of Japan and the use of Agent Orange in Vietnam, acts that killed women, children, and other innocent people. He derided U.S. economic embargoes that, according to him, have also killed children and elderly people in Iraq, Cuba, and other countries for more than thirty-five years. He concluded by calling Americans butchers, liars, and hypocrites.[33]

Whether dealing with terrorists with long traditions or with the new boys on the block, we should pay attention to the advice

from our neighbors in England where terrorism has become a
way of life:

> Terrorists always have a logic, which from their point of view is
> compelling. Indeed, it is so compelling that they have to commit
> the most awful crimes to achieve some greater good. Sometimes
> the logic is as beguiling that you even begin to half believe it, to
> join in with the process and to be carried along by its momentum.[34]

A number of current trends will influence the character of terrorism as it affects Americans. In some ways, politically and economically motivated violence will be more of the same: bombings,
hostage taking, and assassinations, for example. But in other ways
the odds are that terrorist activities will take dramatic turns. In anticipating future threats, a few developments warrant special
attention.

1. *Terrorism and the Media:* Terrorism is at the top of the news
hour, much to the delight of the perpetrators. From the antennas
of CNN and the BBC to studios throughout the world, the media
is often overloaded with both true and false reports featuring
biochemterrorists, pyroterrorists, aeroterrorists, narcoterrorists,
cyberterrorists, and their many cousins.

When airline hijackings became standard fare for terrorists,
prolonged negotiations meant extended air time for the bandits
to plead for political justice and to show their readiness to do battle with even the strongest countries of the world. Within short
order, several countries mobilized SWAT teams, trained and
equipped to descend on the scene and root out the renegades.
Disaffected militants realized that the longer they carried on with
an event, the less the chance of their personal survival. Within a
few years, the era of extended negotiations broadcast to the
world from airports seemed to come to an end. But aircraft hijacking was only one form of terrorism that offered dramatic television visuals.

In 1995, as the Chechen rebels seized civilian hospitals in the
Caucasus mountains of Russia, their negotiations with Russian
political and military leaders took on an immediate sense of
urgency and filled the Russian airways for many days. In Moscow, I sat transfixed, along with the whole city, in front of the television and watched the Sunday morning negotiations between

the Russian prime minister in Moscow and the leader of the Chechen militia 1,000 miles away.

The negotiations followed the Chechen takeover of a hospital where the rebels threatened to kill the sick and the infirm, displaying many of their Russian hostages in the windows of the hospital. In a unique reversal, it was Prime Minister Chernomyrdin who had sought the air time rather than the rebels. He needed to shore up support within Russia for the faltering military effort. He could not have chosen a more effective method of demonstrating to the Russian audience his personal concern for fellow Russians—the safety of the hostages and the plight of Russian soldiers surrounding the hospital who were inadequately equipped, seldom paid, and poorly motivated to pursue the war. The media blitz did much to sway public opinion toward support for his concessions so the rebels would leave the hospital and, soon thereafter, facilitate withdrawal of Russian soldiers from Chechnya.

Across the world in Peru, the hostage odyssey at the Japanese Embassy began in December 1996, lasted for four months, and demonstrated the fickle nature of the media. The presence of diplomats from many countries initially promised many cameras would descend on the compound. But international television interest waned after one or two weeks only to resume months later when the house was stormed, the guerrillas killed, and the hostages released. The terrorists and their followers were surely disappointed that they did not remain headline news, except in Peru and Japan, for a longer period, but the public, at least in the United States, soon went on to other issues with a more direct impact on their lives. Perhaps, had the captors been more bloodthirsty, viewers might have stayed tuned. But the hostages reported that, to the contrary, the guerilla leader was a rather stimulating dinner conversationalist.[35]

The influence of gut-wrenching visuals on government policies can be significant—especially in the country where the events are taking place. Indeed, one of the challenges of counterterrorism efforts is the ability of governments to withstand temptations to change policies in a given situation simply because of public reaction to media coverage. Of course reporters want to be first with the story, in as timely and dramatic a fashion as possible. At the same time the media generally wants to maintain professional standards of accuracy and objectivity. Usually media officials do not object to help, if called upon, in resolving crises if they can do so without losing a story or being exposed to excessive dangers.

Of special interest in this regard are media "codes of conduct" such as adopted by the *Chicago Sun Times* to avoid playing into the hands of terrorists. (See Table 1.)

Several developments are of particular significance for media coverage of terrorism. First, some attacks are designed to be anonymous, as was the case with the World Trade Center bombing. The media can only speculate as to the terrorist's motives but can also stir public anxieties, which may indeed be the primary purpose of the attacks. Second, as terrorism becomes more violent through the use of more powerful bombs, the possibility of public panic in response to real or concocted threats of large-scale destruction increases. Also, attacks on reporters who are outspoken as to the deeds of terrorist groups are on the rise. Dealing with threats of intimidation is not easy, not even for journalists. Finally, American press coverage of U.S. investigations and apprehensions of terrorists abroad may expose details pinpointing countries who have helped us by stretching the limits of their authorities. Such exposure of their roles could dampen their enthusiasm for future cooperation.

Publicity has been described by Margaret Thatcher as the oxygen of terrorism. In return, barring sex scandals, terrorism is increasingly the oxygen of the media, and it must be handled with care.[36]

2. Access to the Weapons Caches of the Former Soviet Union: While the disintegration of the Soviet Union's military machine has triggered Washington's release of funds and energies to combat other

TABLE 1. Media Standards in Covering Terrorist Cases

- Paraphrasing terrorist demands to avoid unbridled propaganda
- Banning participation of reporters in negotiations with terrorists
- Coordinating coverage through editors who are in touch with police
- Providing thoughtful, credible, and restrained coverage
- Allowing only senior editors to withhold or defer information

Source, Raphael F. Perl, Terrorism, the Future, and U.S. Foreign Policy, CRS Issue Brief, Congressional Research Service, February 19, 1997, IB95112, p. CRS-9.

international threats, it also has expanded a full line of weapons potentially available to terrorists. Along with the United States, the Soviet Union had the world's largest arsenal of weapons of all types—from nuclear missiles to Kalashnikov automatic rifles. For decades, the USSR was a major weapons supplier to countries around the world, including countries on America's blacklist—Iraq, Cuba, and North Korea for example. While not a supplier of weapons of mass destruction to these countries, at least not as far as we know, the Soviet Union provided sophisticated electronics, materials, control, propulsion, and other technologies which can be integral components of such weapons. Further, they trained thousands of scientists and engineers from some of these same blacklisted countries in techniques with direct application to weaponry.[37]

Over the period of a decade I have witnessed firsthand the economic disintegration of the support systems that had nurtured the deepest recesses of the military complex of the former Soviet Union. *The leakage of the leftovers of the Soviet military machine to countries which sponsor terrorism and to terrorist organizations is a real threat.* The potential tributaries of this river of resources and know-how are manifold. They could be filled with a variety of items in addition to military hardware—sales and thefts of very dangerous materials (highly enriched uranium, chemical toxins, and biological pathogens, for example), transmittal of technical data on designs of advanced weapons through document exchanges or Internet links, and transfer of weapons expertise by Russian specialists, some truly desperate enough to consider selling their know-how simply to keep food on the table.

A related problem is the legitimate sale around the world of Russian dual-use goods and services that can be diverted from civilian applications to support nuclear, chemical, biological, or other threatening weapons programs of rogue states. Their sale of nuclear reactors to Iran and of sophisticated electronics to Syria, for example, have recently triggered international anxieties.

Also of growing worldwide concern is the reach of Russian criminals, now firmly rooted in every continent, including many with longstanding ties to the former Soviet military and intelligence complexes. They are redefining international organized crime—crime that ruthlessly takes advantage of every opportunity to accumulate wealth. Not only do they trade in weapons and drugs, but they also shake down their own country's prominent ice hockey stars playing in the North American National Hockey League for hefty payments, threatening to harm their families left behind.[38]

3. The Ambiguities of State-Sponsored Terrorism: There are seven countries on the terrorism list of the United States: Cuba, Iran, Iraq, Libya, North Korea, Sudan, and Syria. Several other countries, such as Afghanistan and Pakistan, are on watch lists.

The governments of Iran and Iraq are branded as using terrorism as instruments of policy or warfare beyond their borders. At the other extreme, Cuba and North Korea seem more passive in simply granting safe havens for known terrorists. When discussing global terrorism, the seven governments are just as much terrorist organizations as the dozens of so-called subnational or independent terrorist groups that for decades have provoked worries in the United States. Indeed, some of the independent groups are known to be closely aligned with the seven governments of concern (e.g., Hamas and Hizballah links with Iran).[39]

The recent coupling of governments holding extensive technical resources with terrorist groups accustomed to carrying out assaults on helpless people is indeed a frightening development. This enables these governments to use three overlapping routes in dealing harshly with adversaries: launch a war, use clandestine government agents to attack foreign targets, or simply contract out tasks of destruction to smaller groups.

As we will see, the interests of Iran and Iraq in nuclear, chemical, and biological weaponry add a new dimension to the terrorist threat. While independent groups may eventually develop their own high-tech capabilities, close ties to Iraq, Iran, and possibly other states (which have abundant financial resources to devote to military activities and extensive intelligence networks to provide alerts as to countermeasures) ease the technical problems when trying to use highly destructive devices to attack other countries, including the United States.

In the future, scattered incidents of conventional bombings, poisonings, and shootings aimed at Americans at home or overseas may come from governments hostile to the United States.

When compared to the detonation of a nuclear device in a steamer trunk or the careful release of deadly disease agents from a five-gallon drum, such traditional acts of terrorism can rightly be classified as important but "secondary" threats.

That said, the U.S. government seems to predict that state-sponsored terrorism is not the greatest danger on the horizon and urges the primary focus on so-called independent groups. They argue that states know they can be held accountable and subject to retaliatory steps whereas a more amorphous enemy can be more elusive.[40] As we will see, this view, based on obsolete concepts of terrorism, is shaky at best and is dangerous. Even taking into account the activities of the independent Aum Shinrikyo, I believe that few terrorists will be able to accumulate the funds and armaments needed for serious assaults without some governmental body aiding, abetting, or permitting such activity.

4. *The International Linkages of American Dissidents:* Why would a foreign terrorist group bring guns, explosives, or toxic agents into the United States when they can shop for them right here in many American cities? According to the FBI, international terrorist groups have already set up residence in the United States in close proximity to domestic terrorist organizations. What attracts them is the freedom they have to operate, acknowledges the FBI.[41] According to one report, a Palestinian arrested by Israeli security forces claimed he received bomb training during a pro-Hamas conference in Kansas City.[42] (See Table 2.)

TABLE 2.   Domestic Terrorist Groups of Concern

Right Wing Extremist Groups: Army of Israel, Aryan Nations, Texas Aryan Brotherhood, Our One Supreme Court, Freemen, Republic of Texas.

Militia Groups: California Militia, Viper Militia, Mountaineer Militia, Texas Constitutional Militia, Utah Free Militia, North Idaho Militia.

Puerto Rican Terrorist Groups: EPB-Macheteros, Armed Forces for Puerto Rican National Liberation, Movement of National Liberation.

Special Interest: Animal Liberation Front.

Source: Louis J. Freeh, Prepared Statement, Senate Appropriation Committee, May 13, 1997.

Regretfully, in the United States sufficient funds can buy any type of material or service—including the ingredients for carrying out the most violent type of criminal acts. Criminal elements from Russia, Nigeria, and other parts of the world have demonstrated the ease of amassing financial war chests in the United States that are built on fraud and felony schemes foisted on Americans.[43] With funds in hand, recruitment of willing accomplices in the United States, while risky, can be an attractive one-stop-shopping alternative to importing full-service terrorists from abroad.

5. Drugs, Money Laundering, and Terrorism:

> We intend to keep our counterterrorism efforts separated from counternarcotics programs. We have a chance of making real progress in staying ahead of the terrorists, but in the drug area, we cannot possibly win. Let's not contaminate our successes with our failures.[44]

With these words, a senior Department of State official explained why separate offices handle problems of terrorism and problems of drugs and thugs. "Besides," he added, "drug trafficking is about money, not terrorism."[45]

The days when the economic crimes and money laundering of the drug dealers, the mafia, and the robber barons were not tied to terrorism are over. With hundreds of billions of dollars at stake every year from the drug trade alone, the deadly interlinkages among groups with different agendas but with common tactics of violence are evident around the world. Indeed, the narcotics industry is the greatest generator of terrorism in some countries. Drug profits arm rural guerrillas and urban gangs, promoting corruption that destroys democracy and renders law enforcement agencies impotent.[46] The marriages of the terrorists and traffickers have become daily ceremonies. Arms traders, drug runners, and money launderers will happily support terrorist campaigns in their ignoble pursuit of profit.

Further, those who harbor doubts about the terrorist inclinations of drug lords should pay attention to the comments of a leading Colombian crime figure:

> The drug business is not just money, it is also political. The head of Cali, Gilberto Rodriguez Orejuela, thinks of it as a war in

which he is producing a chemical poison against the United States and its people. Since the U.S. is the only threat to him, he will do all he can to weaken the country.[47]

Whether killing Colombian peasants who resist turning over their farms to drug processors, murdering Bogotan politicians who refuse to be bribed, or dynamiting American-owned oil installations and pipelines, the narcoterrorists of Colombia dramatically symbolize the ties that bind together the narcotics, money laundering, and terrorism industries. Similar examples have existed for years in Peru, Sri Lanka, and Southeast Asia. The trend is toward an expansion of drug-terrorism alliances to other regions as well. These linkages are discussed in more detail later.

To separate measures to combat global terrorism from strategies to stop the narcotraffickers ignores reality and reduces the likelihood of winning either battle. And we cannot forget that worldwide drug trafficking involves hundreds of billions of dollars each year.

6. Megacities as Breeding Grounds for International Terrorists: As I have traveled through poverty areas of many cities of the world over the years—Lagos, Bombay, Bangkok, Karachi, Rio de Janeiro, Djakarta, Mexico City—the potential for violence has become more and more obvious. A former finance minister of Indonesia once reminded me that while improved distribution of income within large cities is a fine policy for any government, there was no income to distribute in his country.[48] In many countries, migrants from the countryside have found despair as they search for a higher standard of living in the urban areas. The megacities of the developing world are becoming incubators for criminals. There, the most aggressive inhabitants hone their survival skills and band together with like-minded individuals.

Within two to three decades there will be more than twenty-five cities with populations exceeding 10 million. The explosive potential of these cities has been described as follows:

This is an environment in which survival takes precedence over the rule of law, in which alienation and anger are rife, and in

which localized street gangs become the dominant form of social organizations. The most ruthless and efficient of these gangs
are likely to develop into more significant and powerful criminal organizations. For some individuals, solace and bonding
will be found in political, social, and economic activism that
could all too easily become transformed into new forms of eco-
nomic-based terrorism. Urban youth provides an excellent recruitment pool for existing transnational criminal organizations,
and megacities are likely to be a breeding ground for a new rad-
icalism rooted in the desire for revenge or simply a fervent wish
to destroy what one cannot hope to have.[49]

Why not make money if you can? Counterterrorism experts
will not solve the problems of population growth, but they can
certainly point out this often forgotten dimension of American
policy in the Third World and trumpet its importance as it relates
to terrorism.

7. The Information Revolution: Computer networking, cellular
phones, and encryption software are terrorist's new tools. Acquiring personal data for forged passports and other official documents amounts to child's play; credit card fraud is a training
exercise for new recruits; and penetrating bank accounts is the
daily fare of the more experienced.

Much more frightening is the damage terrorists might do to
the electronic underpinnings of many American institutions, as
they pump up their budgets to include funds for cybersabotage. For
example, the U.S. Department of Defense uses over two million
computers, with many local, wide-area, and wireless networks. In
1995 there were over 250,000 attempted intrusions, aimed at theft,
alterations, and destruction of software and data.[50] In addition to
electronic trespassing into our financial, power, and transportation
networks, assaults on our police and emergency response systems,
and even on our hospital infrastructures that depend on computer-
based communications may soon become a reality.

It is necessary to add to this array of vulnerable networks
many other sensitive systems of government agencies—at the national, state, and local levels—and those of educational and commercial organizations dependent on computers. Mischievous
short circuits and more serious knots in these electronic webs

could bring the functioning of our society as we now know it to a screeching halt, to the delight of terrorists waiting to pounce on technically paralyzed victims.

Thus, businesses and government offices around the world will soon be on constant alert, spending billions of dollars to ensure the integrity of their bookkeeping and inventory control systems, transaction billings and payments, and proprietary information and personnel records.[51]

For many U.S. government agencies, international terrorism has replaced Soviet nuclear-tipped missiles as the major national security concern on the world stage. As the federal budget for countering Soviet nuclear capabilities declined, these agencies quickly identified terrorism as a growth area that could easily absorb the long-awaited peace dividend. Using military jargon, several parallels may be drawn between the perceived Cold War dangers and those currently posed by terrorists: within a specific timeframe the probability of an attack is unknown, the severity of attack is unpredictable, and the list of targets is assumed but not verifiable.

These dueling threats, in fact, are not quite parallel. Unlike what was a low probability of a Soviet attack over the years, there stands little doubt about the likelihood of terrorist attacks on American soil now and in the future. Also, Cold War targets of nuclear weapons were confined largely to major population centers, military bases, and transportation and communication hubs,

TABLE 3.   Examples of Threats to Computer Systems

- Incompetent users, inquisitive users, unintentional blunders.
- Hackers driven by technical challenge.
- Disgruntled employees or customers seeking revenge.
- Crooks interested in personal financial gain or stealing
- Organized crime interested in financial gain or covering criminal activity
- Terrorist groups or nation-states trying to influence U.S. policy by staging isolated attacks.
- Foreign espionage agents seeking to exploit information for economic, political, or military purposes.

Source: *Report of the Defense Science Board Task Force on Information Warfare Defense,* Defense Science Board, Department of Defense, Washington, November, 1996, p. A3.

but the targets of terrorists range from large arenas of people at sporting events to individual buildings, to bridges, to computer systems—anywhere in the country. In the end, we know terrorist attacks—on American citizens, on American facilities, and on government and financial institutions—are inescapable. Though death and destruction from any individual attack will be less than the devastation wrought by a Soviet missile, the random nature of the targets can cause an enormous wave of public fright.

As an FBI spokesperson asserted, *terrorism has replaced the Cold War as our battlefield.*[52] The FBI and other agencies are now implementing this philosophical shift. The Central Intelligence Agency (CIA), which devoted huge resources to uncovering Soviet secrets, has found new directions through its recently established centers on counterterrorism, counterproliferation, and counter-organized crime. And dozens of units of the Pentagon have quickly reinvented themselves, initiating programs and strike forces designed to counter global terrorism.

In short, filling the post-Cold War void has provided new opportunities to accelerate U.S. antiterrorism policy that began to take shape after the 1972 Olympics. During most of that time, our narrow vision was directed primarily to Cold War dangers and did not adequately recognize broader global threats. Now, the disintegration of the Soviet Union, rather than lifting the curtain on a new dawn of world peace, has flipped the mental remote control devices to some very unsavory images of weapons proliferation intertwined with international organized crime, corruption, and politically inspired violence aimed at Americans.[53]

Despite the difficulty in pinning down an evolving term, academics and practitioners must continue their never-ending struggle to define terrorism. Of special relevance is the debate over particular categories of laws and punishments for those who carry out activities classified as terrorism. How do we distinguish whether the use of a gun or a gas canister should be classified as domestic crime, urban crime, or terrorism? Motivation is the key say the experts; and motivation to shock, stun, or intimidate a target group wider than the immediate victims is an important test, they add.[54]

3. The cast of characters from conventional terrorism to superterrorism: Carlos the Jackal, infamous freelance terrorist of the 1970s and 1980s; Shoko Asahara, Aum Shinrikyo cult leader responsible for nerve gas attack on the Tokyo subway in March 1995; and British computer hacker Richard Pryce who broke into computer systems operated by the U.S. Air Force and the Lockheed Martin Corporation.

Confusing things further is the fact that both states and independent groups are active on the international stage. I have noted that it may be easier to punish governments that promote terrorism than to deal with splinter groups that are more difficult to define, much less to locate. The most challenging of all is to cope with unholy alliances of rogue states and militant groups in the service of these states.

No matter how we choose to define or characterize terrorism, we cannot underestimate its power and inevitability. Indeed, current definitions of terrorism fail to capture the magnitude of the problem. It is a real and growing threat to the peace and stability of the world—with the front lines of conflict slowly encroaching on the sanctity of American neighborhoods.

No combination of affordable measures will provide complete protection—just as powerful armies do not deter all wars, the best police departments will not stop all crimes, and the most benevolent governments cannot satisfy all constituents or avoid all violent protests.[55] However, just as we continue to raise the level of sophistication in the way we arm and direct our military forces, train our law enforcement agencies, and organize government programs, so must we forge ahead with new initiatives that combine these capabilities to thwart terrorism. By using more aggressive means to interdict terrorists at early stages, by expediting prosecution and more severe punishment of terrorists, by establishing effective and synergistic partnerships among the legal and law enforcement systems throughout the world, and by leading a long overdue assault on at least some of the root causes of terrorism (despite the pessimism of Walter Lacqueur and others), we should be able, in time, to reduce the threat of this modern-day sword of Damocles poised over the United States and the rest of the world.

The Nuclear Legacy: A Shopping Mall for Rogue States and Terrorists

Someday, possibly quite soon, the world will wake up to face a rogue regime or a bunch of terrorists with a nuclear bomb and the will to use it.
The Economist, 1997

Three-quarters of informed Americans find it likely that the United States could be attacked by terrorist groups that smuggle nuclear bombs into the country and set them off.
Opinion Survey of the
Mellman Group, 1997

We should be like the Chinese—poor and riding donkeys, but respected and possessing an atom bomb.
Muammar Qaddafi, 1987

*A*ll was still at the Sevmorput naval shipyard at Murmansk, Russia. In the darkness a captain from the naval base, Alexei Tikhomirov, quietly made his way through a huge hole that formed a ragged entrance in the fence surrounding a nuclear fuel storage area. He walked briskly over to a nearby storage building. He whipped out a small saw and cut through a padlock before prying open the battered wooden door. Once inside, Tikhomirov carefully removed several pieces of steely-gray metal from an unused, forsaken fuel rod for a nuclear reactor. These bits of seemingly unexceptional metal, which he stuffed into a bag, were in fact 4.5 kilograms of highly enriched uranium, called *HEU* by nuclear specialists. Retracing his steps, Tikhomirov climbed into the car where his accomplice, fellow officer Oleg Baranov, waited. Baranov dropped Tikhomirov at his house. Then Baranov drove to his own home where he stashed the bounty in his garage until arrangements could be made to sell it for desperately needed cash.

Twelve hours passed before the incident was discovered. The theft was revealed only because, in his haste, Tikhomirov had left the back door of the storage building open. The guards had un-

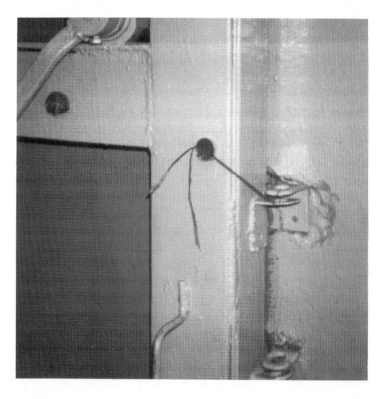

4. A Soviet uranium storage facility, secured with string and sealing wax.

covered the theft in a routine search of the area. Had Tikhomirov not been so careless, the theft might have gone unnoticed to this day. Six months later, the details of the plot came to light. Baranov's brother spilled the beans in his effort to market the cache. He asked yet another naval officer for help in finding a buyer for the stolen uranium. This officer, resisting the lure of easy rubles, reported the theft to local authorities who promptly arrested the triumvirate and recovered the uranium.[1]

The local investigator of this incident at a major repository of fuel for nuclear submarines described security at the facility as follows:

> On the side of the shipyard facing Kola Bay, there is no fence at all. You could take a dinghy, sail right in—especially at night—and do whatever you wanted to do. On the side facing the Murmansk industrial zone there are holes in the fences everywhere. And even those places where there aren't holes, any child could knock over the half-rotten wooden fence boards. Potatoes were guarded better.[2]

This incident was reported by the press in both Russia and the West in late 1994. The uranium heist gone awry exemplifies the consequences of the shoddy security practices that have become all too common in the former Soviet Union.

How can such an appalling situation be possible? After all, in Soviet times, security was an integral part of the nuclear environment. Highly disciplined armed guards with top-level clearances and unfettered access to the most sensitive facilities conducted constant surveillance under the watchful eyes of security agencies.

The relaxation of the political system, once notorious for its rigid strictures concerning the surveillance of people and materials belonging to the state, together with severely limited budgets particularly at government facilities, have resulted in fewer guards. Moreover, the remaining poorly paid guard positions are not the easiest to fill with committed professionals. Private security agencies have become one of the biggest and most lucrative growth industries in the former Soviet Union. Reports indicate that Russia's new capitalists have spent hundreds of millions of dollars each year for protection. While investing in armor-plated cars, bomb sensors, hidden cameras, bulletproof vests, antiwiretapping gear, and thousands of weapons, they have recruited many a bull-necked veteran of the Afghan and Chechen wars as bodyguards.[3] With this new demand driving up the price of guard service, underfinanced government institutions suffer high turnover and low quality in security personnel, an area at one time considered a top priority.

As to other security measures, unlike the United States where formidable physical barriers and stringent accounting systems thwart illicit efforts to divert nuclear materials, the Soviets always gave such protective measures far less priority than deterrence by guards. Besides, Russian managers now find themselves with almost no funds to maintain the flimsy barriers and fences that are in place, let alone funds for upgrading protection systems to international standards. And they lack both skills and incentives to install modern accounting systems.

Ironically, these factors make the nuclear stockpile even more sinister than when it was under Soviet lock and key. With the possibility of new bridges linking corrupt officials, desperate facility managers, and international crime syndicates, the nuclear cache could become a modern day Pandora's box. Such linkages may already exist. For example, in December 1994 police in Prague responded to a phone tip and seized 2.7 kilos of *HEU* from the back seat of a car. The material, presumably originating from the former Soviet Union, had been lifted by a trio of former employees of the nuclear industry: a Czech, a Belarusan, and a Ukrainian. Though never confirmed, the heist could have been part of an international crime operation.[4]

The former Soviet Union's nuclear weapons arsenal stretches across eleven time zones from the border of Poland to the Pacific Ocean. *Tens of thousands of nuclear warheads*—deployed in missiles, maintained at storage sites, and held in various stages of assembly and disassembly—*are scattered at perhaps eighty locations across Russia*. The containment challenge is awesome. The United States began reducing its smaller and more compact inventory a few years ago. Our own more modest problems of responsible dismantlement have proven ponderous and expensive even for our thriving economy.

During the fall of 1997, millions of Americans were alerted to the easy access to the Russian nuclear warehouse. The release of the fictionalized film *The Peacemaker*, featuring theft of a Russian nuclear warhead that could be carried in a backpack, coincided with televised statements by Russian general-turned-politician Alexander Lebed that Russia could not account for 100 nuclear suitcase bombs assembled during the Soviet era. The filmmakers claimed their movie was based in large part on a nonfiction book devoted to exposing poorly guarded nuclear weapons in Russia. Lebed contended that each missing sabotage suitcase could destroy a population of 100,000 or more. These bombs were left behind in the Baltic countries, the Caucasus, and the Ukraine when the Soviet army pulled back from these areas a few years earlier.[5]

Some American experts denounced the film for raising concerns unnecessarily, accusing the filmmakers of stimulating the growth of a new cottage industry of ambulance chasers of loose nukes. Others, however, were pleased that despite the film's shortcomings, it provided a wake-up call for the serious problem of potential theft of the most dangerous explosives and related materials.[6]

While seemingly authoritative sources in both Russia and Washington immediately debunked Lebed's claim that portable nuclear devices were unaccounted for, the Pentagon acknowledged small sabotage devices had been built in both countries. Initially, the Pentagon suggested these weapons weighed between 100 and 200 pounds and could be carried in backpacks only by "very strong men."[7] However, by the end of the year, the Secretary of Energy admitted that two decades earlier the United States had constructed devices weighing only sixty pounds that were to be delivered by parachute, particularly in harbor areas. These were in addition to 160-pound weapons which were to be carried and detonated by Green Berets or Navy SEALS.[8] It must be assumed the Soviet Union had developed comparable weapons, which leds at least limited credibility to Lebed's claim.

The Peacemaker did indeed highlight one of the weakest links in Russia's nuclear security system—protecting weapons during transport by train across barren areas of Siberia. Scene of the fictionalized theft, this vast stretch seems the perfect backdrop for organized criminals wishing to encounter few impediments. At the same time, Lebed raised an important question as to the reliability of Soviet and now Russian security measures. Given the chaos that transpired during the breakup of the Soviet Union, it seems nearly impossible to prove every Soviet-built warhead is in safekeeping.

If the threat needs further underscoring, the CIA frequently quotes a "knowledgeable Russian source" as follows:

> Accounting procedures are so inadequate that an officer with access could remove a warhead, replace it with a readily available training dummy, and authorities might not discover the switch for as long as six months.[9]

The agency adds that nuclear weapons handlers, who used to be among the best-treated soldiers, now face a crisis situation in housing, pay, food, manning levels, and social services. This change has resulted in plummeting morale and lapses in discipline.[10]

I heard firsthand of the troubles of one of the former elite in mid-1997 while visiting a Russian friend. As an unskilled worker, her own economic situation had rapidly deteriorated. Her several part-time jobs supplied multiple small paychecks adding up to about $80 per month, just enough to cover rent and utilities. Yet, her brother phoned her for a loan. An army colonel stationed at a sensitive facility north of Moscow, he had not received a paycheck for six months. His only source of income was the produce from his dacha garden.

Even the Russian Minister of Defense, in lamenting the "horrifying" state of the Russian armed forces, reported in 1997 that Russia might soon reach the threshold beyond which its missiles and nuclear systems could not be controlled. He added that military reform was making no progress since the effort was sorely lacking in political will from senior levels of command, the ministry had failed to formulate a general strategy, and consistency of policies and action had yet to be demonstrated. The minister went on to worry aloud about renegade officers firing weapons without authorization or, more likely in his view, conspiring to sell military assets to bidders from abroad.[11] The fact that suicide rates among military officers was on the rise further underscored his concerns.

The concept of military professionalism had already suffered a near-fatal blow during an era of military stupidity in Chechnya. Now, only through harsh disciplinary measures is a semblance of

order maintained; such discipline is often manifested in harass-
ment of new recruits. Further undermining their self-esteem, the
Russian press feels no compunction publishing photos of soldiers
picking cabbages to prevent their rotting in the field. Army paychecks, if received at all, cover only the costs of a few packets of
cigarettes. With uniforms often in tatters, soldiers, sailors, and fly-
ers—like most of the nation's population—resort to whatever
methods they can to support their families.[12]

Of special concern are the problems associated with demobi-
lization of Russian officers previously assigned to missile bases in
remote areas of Siberia. Many of the missiles at facilities they once
guarded have been dismantled as a result of arms control agreements limiting the size of the strategic rocket forces. The warheads
are all but abandoned at temporary storage sites in isolated locations. Former officers, legally entitled to housing and pensions
upon discharge from service, live in tent cities pitched in close
proximity to their former workplaces and plant vegetables to survive. These retired officers, armed with vital knowledge relative
to the security of the nearby nuclear weapons they once helped
guard—and the cynicism fueled by unfulfilled promises—could
represent an incendiary mix. No wonder the CIA gives credence
to the seemingly unbelievable stories from Russian sources with
ties to the military.

That said, the statement of the chief of the U.S. Strategic Air
Command in October 1997—that he had traveled to Russia to dis-
pel perceptions that safety and security of Russian nuclear
weapons were at some risk—was very surprising. When he announced that military downsizing was going well within the
Russian Strategic Rocket Forces and there was absolutely no indication of unrest, my assumption was that he had been led astray
by overly generous and optimistic hosts.[13]

The U.S. government has taken a number of steps to assist
Russia in minimizing dangerous side effects of the downsizing of
its military forces. The most ambitious efforts were triggered by
passage of important legislation introduced in 1991 by Senator
Richard Lugar (R-Indiana) and former Senator Sam Nunn (D-
Georgia). Senator Pete Domenici (R-New Mexico) has recently
added support to their efforts. Cooperative programs supported
by American funds and expertise have emphasized the upgrading

of physical security surrounding weapon sites, protection of excess warheads during their storage and shipment to dismantlement sites, and improvement of accounting procedures. While limited in size, these programs have stimulated a sense of urgency among some Russian authorities.[14]

Still, the possibility exists that desperate, corrupt, or coerced officials with authorized access to weapons storage facilities could orchestrate thefts of warheads or their components for the right fee. Only a personal code of integrity among professional security personnel that rebuffs procedural shortcuts for securing weaponry will improve international confidence that stockpiles are in good hands. Such a code of professionalism is closely tied to faster economic recovery and to better control of organized crime. In the meantime, vigorous support for the Nunn–Lugar–Domenici programs can apply some of the necessary bandages while waiting for more enduring fixes to a problem of such critical importance.

In addition to the nuclear payloads already encapsulated in warheads and in warhead components, Russia retains between 600–700 tons of *HEU* and plutonium.[15] These requisite ingredients for weapons of mass destruction are housed at research institutes, educational institutions, nuclear reactor sites, transportation depots, and in warehouses and storage yards, including the Sevmorput naval shipyard referred to earlier. The material may be found in bars, scrap, reactor fuel rods, powder, or other configurations, illustrating the multiple uses for such energy-intensive materials. Aside from their military-related uses, these materials can be directed to a variety of productive tasks from fueling electric power and civilian research reactors to providing the small power sources for pacemaker devices.

Nuclear weapons have been constructed in the United States and in other highly advanced countries using only a few kilograms of plutonium or several times that amount of *HEU*. The amount of such material required depends on the type and sophistication of the device; even if fifty kilograms of material are needed for a very crude device, that quantity represents a minuscule amount compared to the many tons languishing in poorly secured buildings throughout Russia.[16]

Making matters worse, as former CIA Director John Deutch points out, Russian officials do not really know how much *HEU* and plutonium they have.[17] For decades, the accounting techniques used in the Soviet era were intentionally fluid. Much of the material was kept off the books so that managers would have flexibility when reporting their successes in fulfilling monthly quotas to produce such material. Now, new managers who inherited poorly quantified inventories have even fewer resources than in the past to devote to accounting systems and practices. Thus, both government ministries and enterprise directors have great difficulty in providing precise information on the size and contents of their inventories.[18]

While never invited to tour a Russian weapons site, I have visited numerous institutes and enterprises of the former Soviet Union that continue to be repositories of *HEU* and plutonium. Like other American visitors from government agencies and laboratories, I've been sensitized to the U.S. focus on the importance of stringent control over nuclear material. Thus, the inadequate attention paid by the Russian nuclear leadership to the physical security of such material is highly unsettling.[19]

In 1996, while visiting a major Russian research facility I learned that hundreds of kilograms of *HEU* were stored in a warehouse with flimsy wooden doors that had slipped off their guide rails. Two guards were said to be inside the building, the major storage site for the facility. One additional soldier was visible at an outside guard post. This was quite a contrast to American storage facilities: specially designed concrete buildings behind sturdy fences, surrounded by remote sensors. While three guards would have seemed adequate if the facility were inherently secure and difficult to penetrate, physical ruggedness was not an attribute of this building with its warped doors vulnerable even to the effects of heavy rain.

Most surprising, the uranium warehouse was located immediately adjacent to a semi-enclosed building stuffed with paper products belonging to a Western firm. Shifts of truck drivers with no visible supervision were loading and unloading cardboard boxes from inside the building. Then, without being inspected, they drove their trucks through the gate and off the site. Surreptitious loading of uranium bars into the cardboard cartons seemed like a piece of cake.[20]

The warehouse was but one of twenty storage locations remaining at the institute—a four-fold reduction from the eighty on site several months earlier. To their credit, Russian officials had orchestrated a crash program to consolidate material, after an inspection had revealed numerous security problems. However, this porous warehouse was advertised as the most secure of the twenty buildings. One can only wonder what the others were like.

Uranium transportation from the facility raised additional concerns. No armored trucks or specially designed vehicles were assigned to this institute. I was told that each week uranium in some form was loaded onto ordinary trucks and driven to other institutes within a radius of 50–100 miles. This casual system contrasts sharply to the special transportation vehicles and extensive procedures used in the United States for tracking the fate of each gram of nuclear material.[21]

During the early 1990s, the Russian government was reluctant to open the gates to their previously secret facilities despite the eagerness of Americans and other outsiders to cooperate in tackling the storage problem. They hesitated even though, in the absence of financing from abroad, they were in no position to guarantee the security of the dangerous remnants of the Cold War. Finally, after acknowledging a number of thefts and attempted burglaries of *HEU* at their facilities, the Russian nuclear authorities agreed to programs of cooperation that would upgrade the physical protection, control, and accountability of nuclear material at several dozen sites across the country.

Against this backdrop, the U.S. Department of Energy has spearheaded international efforts to help contain nuclear materials *within* Russia and the other states of the former Soviet Union. The U.S. Congress has poured hundreds of millions of dollars into support programs for American specialists to work with Russian security experts in upgrading the protection and accounting systems at many sites where *HEU* and plutonium are stored. Much of the money has been funneled immediately into Russian institutions, because they lack sufficient funds to cover even the salaries of their specialists who participate in collaborative activities.[22]

By 1997, the Department of Energy began to announce the success of the cooperative efforts. They reported, for example, that American specialists were working in cooperative programs at about 90 percent of the fifty sites where *HEU* and plutonium were stored in Russia. However, the task of upgrading to international standards the security systems in several hundred buildings scattered across the Urals and Siberia, as well as around Moscow and St. Petersburg, is nothing short of daunting.

The Department of Energy's target date for reaching an acceptable level of security at all of the sites is 2002, but in 1997 the scorecard was as follows: *tons* of *HEU* and plutonium were within sufficiently secure systems; *tens of tons* were within partially secure systems; and *hundreds of tons* remained outside secure systems. Since the remnants of the old systems are far from trustworthy in the prevention of major losses of material, foreign experts must continue to pressure the Russians to maintain the momentum in upgrading technical controls.[23]

Problems of poorly guarded material have not been limited to Russia but have, at times, been more worrisome in other successor states of the former Soviet Union. Fortunately, the amount of *HEU* and plutonium remaining outside Russia is very small. As of 1997, levels are estimated at less than one-half ton compared to the many hundreds of tons in Russia. Most of this material is in Ukraine and Kazakhstan, but there are other locations as well. A review of several sites should delineate more clearly the overall problem.

• *Unveiling Uranium in Kazakhstan:* In 1994, the United States removed more than one-half ton of *HEU* from Kazakhstan in a secret airlift operation called Project Sapphire. The Kazakh government had inherited the *HEU* from the Soviet defense industry. When nuclear testing and other weapons-related activities ceased in Kazakhstan, there was no foreseeable use for the material. The Kazakhstanis convinced U.S. officials that the potential theft of the *HEU* represented an international security threat and that the best way to prevent such a possibility was for the United States to make a bulk purchase and remove the *HEU*. This was accomplished through a cloak-and-dagger air lift from Kazakhstan to Oak Ridge, Tennessee. For their part in providing manpower as well as the uranium, the Kazakhstanis eventually received $3 million from the United States and a significant amount of technical assistance to help with conversion of military facilities to peaceful purposes.[24]

For more than a year, the operation was blanketed in secrecy, at least within Washington. The reason for secrecy was unclear to many Kazakhstanis and to the Russians, who were also parties to the arrangement, but they went along with American insistence on secrecy. Why would the Americans insist on a blackout on such an admirable project that effectively addressed proliferation concerns, economic needs of Kazakhstan, and political tensions among the three countries regarding nuclear uncertainties? Think about it. Would the Iranians get wind of the operation and make a better offer to Kazakh officials? Would the American environmentalists block the transport of *HEU* into Tennessee? Would some members of the U.S. Congress object to using U.S. funds to solve a Russian–Kazakhstani problem if they were aware of the airlift before it was completed? There is some truth in each of these presumptions. In addition, U.S. officials always seem compelled to put high-level security cloaks over all things nuclear.[25]

• *Reducing Capability for Nuclear Research in Georgia:* In 1994, another incident came to light in the Republic of Georgia, 1,000 miles to the south of Moscow. I learned of this development following my discussions with President Shevardnadze to establish cooperative programs through the International Science and Technology Center in Moscow, which I directed at the time. In 1993, after rebel forces had occupied Abkhazia (a province of Georgia along the coast of the Black Sea), a former Soviet nuclear research center near the provincial capital of Sukhumi was closed and the scientists fled to the safety of the distant Georgian capital of Tbilisi. A caretaker crew remained at the center to try to keep strangers out of the complex where several kilograms of *HEU* had been hurriedly placed in an underground vault.

During a subsequent trip to Georgia in 1996, I learned that "professional thieves"—an obvious reference to Abkhazian rebels—broke into the complex earlier that year and managed to extract an undisclosed amount of the *HEU* from the vault. They smuggled their booty into Russia, planning for it to reach western Europe, so the story went. After their long transit across Russia, perhaps hiding the *HEU* in their underwear as done in other smuggling efforts, the thieves reportedly were intercepted by Polish border guards armed with geiger counters. The rest of the story (including the disposition of the material and the fate of the smugglers) has never been revealed. Indeed, even the portion of the story noted above has not been reported to the public.[26]

During that same trip, I heard rumblings of diplomatic disputes centered on another old nuclear research complex near the capital city of Tbilisi. I contacted the director of the Institute of Physics, and within an hour, was seated in the director's car heading to the research reactor site just outside the city. The dismantled reactor was located in an unoccupied complex of about ten buildings, with three uniformed guards on duty at the entrance gate. They received us as though glad for the company. After providing an enthusiastic briefing as to how their alarm siren atop the guard building would summon colleagues from a nearby encampment in an emergency, they quickly arranged to open the buildings.[27]

The principal interest of the Georgian scientists who led the tour of their facility was to secure funding to restart the reactor. The containment vessel and the electrical control system needed replacement—at a cost of several million dollars. They were convinced that revival of Georgian nuclear physics was critical if Georgia was to regain its rightful place as a fertile field for training world-class scientists.

They were proud of their new security system to protect the five kilograms of *HEU* and a small amount of plutonium left over from the reactor's active days. Earlier in the year, the International Atomic Energy Agency (IAEA) and the U.S. government had assisted in the installation of a rather crude barrier consisting of several tons of bricks piled against the door leading to the small room where the material was stored. Joining the bricks in the anteroom were a television monitor and a motion detector, each linked to the guard post with the siren I had seen earlier. Squirrels regularly raced through the anteroom setting off the siren, thus confirming that the detector worked.

Despite these safeguards, the U.S. government was soon insisting that the material be sent to Russia. For Georgian acquiescence to this latest demand, the bottom line was reimbursement for the cost of packing and shipping the material. Diplomatic discussions between American, Georgian, and Russian officials highlighted the constraints posed by Russian environmental laws and the problems of appropriate radioactive shipment containers. But the Georgians were convinced the delay was due to scarce funds. They wanted to remove the headache, but they needed $500 thousand to pay for packing and shipping the material to Russia. Of course, if the United States would also pay them the estimated value of the material—$100 thousand—such an incentive would no doubt raise their enthusiasm for the transfer. In early 1998, the United States and Great Britain finally arranged for the material to be taken to Scotland where it was to be processed into a form useful for research.[28]

An important aspect of nuclear proliferation is the very real possibility that Kazakhstan, Georgia, and other countries surrounding Russia could become favorable routes for nuclear smuggling—through the deserts, over the mountains, and along the coastlines. Countermeasures must continue to emphasize containment of material *at the source* in Russia and other countries. Once the material is illegally and surreptitiously removed from a storage area, the difficulty of intercepting it en route to foreign buyers increases dramatically. Nevertheless, interdiction efforts at international borders, between the original storage areas and the borders, and across transit countries can greatly complicate efforts, making life difficult for contraband mercenaries. Hence a *layered defense*, with strategies directed at every possible checkpoint in the trafficking routes of *HEU* and plutonium, is an essential ingredient of any program to secure nuclear materials. The local intelligence services and the customs services are critical in pinpointing weaknesses in the containment rings and in focusing the interdiction efforts. Of course, U.S. support for these services in Russia and in the other states of the former Soviet Union is a delicate political matter, but given its importance, such support needs to be pursued and maintained.

Even with increased security efforts, loopholes will persist. A Russian specialist in export control described one such gaping hole to me during a trip to Moscow in 1996. While we were discussing the possibility of nuclear trafficking through Central Asia, the specialist promptly pointed out that the simplest and safest way for the Russian mafia to export from Russia any valuable item, be it uranium or gold bullion, was simply to route the contraband through one of the military airports surrounding Moscow. These airports are off limits to the Customs Service and other law enforcement authorities and have their own rules for controlling shipments loaded onto military aircraft bound for destinations throughout Russia and outside the country. This highlights the ease of foiling responsible law enforcement efforts when all departments and levels of government are not coordinating their efforts.[29]

Perhaps the best ammunition for countering nuclear trafficking in the former Soviet Union is simply "showing the money" to informants. Offering generous bounties on the heads of nuclear thieves should be considered not only by the Russian authorities but by the U.S. and other governments concerned about nuclear trafficking. The going rate in 1997 for hired assassins in the region was only $5 thousand per hit. Thus, rewards in the tens of thousands of dollars for information that would stop smugglers before it's too late, for example, should surely attract the attention of many potential whistle blowers, even at Russian military airports. The costs for paying informants would be negligible in comparison to rewards often offered after the fact and to the costs many countries incur in trying to protect themselves from successful diversions of nuclear material.[30]

The principal logistical hurdle faced by rogue states or terrorist groups bent on developing their own nuclear devices, however primitive, is the acquisition of *HEU* or plutonium. Thus, it is essential to safeguard these two materials. In the near term, *HEU* is likely to be the material of choice since the construction of a device which uses *HEU* is much simpler than that of a plutonium device. Further, if international renegades choose to acquire their nuclear materials through theft and international smuggling, they will no doubt factor in that *HEU* is easier to conceal in transit than plutonium which emits radiation that is easier to detect at border checkpoints.[31]

The rogue states of primary concern have in recent years concentrated on developing their own capacity to produce *HEU* or plutonium, apparently assuming that the theft of an intact weapon or of weapons-grade nuclear material was simply too difficult. Production has required large facilities that are difficult to hide.[32] If *HEU* or plutonium were to become available on the international black market, the situation would change dramatically. *Some Western experts believe that sufficient HEU has already been removed from the former Soviet Union to create an active international black market.*[33] The relative simplicity of using easily disguised warehouses to stockpile small quantities of stolen material until enough is accumulated to manufacture a device is an attractive alternative to producing such material—and thus has not been lost on terrorist-oriented governments.

I cannot confirm the contention that significant amounts of *HEU* are available on the black market. It is far from certain that the leaders of the crime syndicates that control the black market have a great interest in the rather unpredictable trade in nuclear material, particularly when they have so many other lucrative enterprises. But we cannot afford to deny the possibility either. The international law enforcement community should increase surveillance, assert interdiction of suspected activities, and more vigorously prosecute nuclear-related crimes to shut off this avenue to would-be thieves. Diplomatic efforts (and in extreme cases even military actions) should be mounted to thwart any efforts by hostile countries to import or produce *HEU* or plutonium for illicit purposes.

• *Nuclear Players and Aspiring Understudies:* As we have noted, of the nuclear weapons states, Russia heads the list as being a potential source of technology and materials that could benefit terrorist operations. In addition, China is regularly criticized by the United States and others for selling very sensitive nuclear equipment to Pakistan and Iran. China undoubtedly views its links with Pakistan as an important balance to India's powerful program and seeks a presence in the Persian Gulf region in view of its dependence on foreign sources of oil to fuel its expanding economy. While China has declared it will not provide assistance at any foreign facilities not subject to international inspections, the U.S. government is nevertheless quite unsettled by its dealings with these two countries.[34]

A brief review of the nuclear ambitions of several countries illustrates the many twists and turns in a plot that suggests we'd better learn to cope with the possibility of nuclear terrorism.

• *The Once and Future Iraq:* In recent years, Iraq has been one of the leading nuclear renegades. "Had Saddam Hussein delayed the invasion of Kuwait by six months, he could have had a nuclear device at his disposal,"[35] said an American weapons design expert, who participated in UN inspections of Iraqi nuclear capabilities following the Persian Gulf War. He noted that immediately after the invasion began, the Iraqis undertook a six-month crash program to divert uranium from a peaceful nuclear research reactor. At a secret centrifuge facility, they worked to enrich uranium to a level suitable for use in a nuclear warhead. He was confident that had it not been for the Iraqi surrender, the program would have succeeded.[36]

Commenting on the future capabilities of Iraq, he recalled his encounter in Baghdad as far back as 1995 with leading Iraqi weapons designers as follows:

> When the UN team expressed doubt that the Iraqi engineers had really destroyed their metallic molds for shaping a critical component of a nuclear warhead as required by the UN resolution, the Iraqis replied that they certainly had destroyed the molds. However, the Iraqis added that if the foreign inspectors would feel better, Iraqi engineers would quickly produce more molds and let the inspectors witness their destruction as well.[37]

Why Saddam Hussein rushed the invasion of Kuwait before detonating a nuclear device is a topic for speculation by military historians. The consequences of Iraq's exploding a nuclear weapon prior to or during the war would surely have changed the course of history in the region. Even a test of a device in the desert would have sent intimidation waves throughout the region. Since no one would have known whether Iraq's inventory was limited to a single test device, the decision as to an appropriate international response would have been difficult.

We now know that the Iraqi nuclear program at one time involved fifteen thousand people. They worked at more than twenty sites at a cost of $10 billion over a period of more than a decade—a decade when the country had not yet been branded an international outlaw. Technical assistance and equipment from Western-based companies and the lax export control regimes of European countries contributed significantly to the Iraqi effort, particularly with regard to the enrichment of uranium.[38]

The Iraqi plan, which may still be on the books, was to place nuclear weapons in missiles capable of penetrating the air defenses of several of the country's neighbors. They had parallel schemes concerning chemical and biological warheads. In considering such a plan, they recognized the need to overcome American military detection capabilities that would come into play. The specific technical challenge was to modify Scud systems obtained from the Soviet Union, so they could carry larger warheads than anticipated in the original designs of the Scuds. As for the guidance capability for the missiles, they reportedly enlisted Russian intermediaries to scavenge components from Soviet systems dismantled because of the arms control agreements.[39] At the same time, the Iraqis needed to reduce the size of the primitive warheads they had already developed. They were on the track to solving both problems when the Allied military coalition began to shut down their program.

The Iraqi nuclear program had already received a severe setback a decade earlier when Israel bombed the principal research facility in Baghdad. But it recovered. And despite any short-term Western squelching of the current threat, in the long term, an oil-rich Iraq with pent-up hatred for its neighbors will surely attempt to revive and expand its most recent nuclear efforts.

At the end of 1997, the IAEA reported that the Iraqi nuclear program was in a dormant state.[40] In the subsequent furor over Iraq's denial of access by inspectors of the United Nations Special Commission (UNSCOM) to the presidential palaces of Saddam Hussein, neither the United States nor any other government publicly suggested that Iraq was concealing a clandestine nuclear program, rather the U.S. has concentrated its ire on hidden chemical and biological weapons. However, with the recurring game of cat and mouse between Saddam and the inspectors, it is sometimes difficult to feel any real confidence that the nuclear program is dormant, let alone dismantled.

In looking ahead to preventive measures which might be taken by the international community in the Middle East and elsewhere, it is instructive to examine why the United States and other Western countries were caught by surprise by the nuclear build-up in Iraq. The following critique by a leading American weapons designer who spent many years monitoring activities in Iraq seems crucially important to future prevention:

> There can be no question that the major share of the responsibility for the failures and weaknesses of the IAEA that led to its not detecting anything being amiss in Iraq's peaceful nuclear program prior to the Gulf War must be borne by the United States, and specifically by its representatives to the IAEA's Board of Governors. The United States, because of its technical and financial leadership in establishing and monitoring the development of international safeguards, was in the best of all positions to understand the weaknesses of this system and the lack of efficient management of it by the IAEA. Prior to the Gulf War, however, there is no record of the U.S. government providing any substantial critique of the safeguards system or taking a leadership role in challenging the system to improve its capability to detect clandestine programs. In fact, the record of U.S. participation in the IAEA in the decade prior to Iraq's invasion of Kuwait was very much of a defensive "don't rock the boat" character.[41]

While coalition building was critical to the success of Desert Storm, it was American leadership that made the operation possible. Similarly, while concurrence of the IAEA Board of Governors on inspection procedures may be important (keeping in mind that Iran, Iraq, Libya, and India have in the past served on the board), the United States cannot simply rely on the IAEA as a counter to the threat of nuclear intimidation by terrorist states or groups. American intelligence possesses unique capabilities—reconnaissance satellites, electronic eavesdropping systems, and connections with the Israeli intelligence service, for example—and should take the lead in ensuring that surprises do not become commonplace. One more surprise could be disastrous.

• *Crisis-Ridden North Korea:* North Korea is an aloof country of closely-guarded secrets. Five decades ago, Soviet military authorities planted the seeds of the nuclear weapons program in North Korea while extracting uranium ores in exchange for Soviet military equipment. Within a few years, North Korean nuclear scientists began to travel to the Soviet Union for advanced training.

By 1970, drawing on Soviet skill and eventually on communist Chinese expertise, North Korean capabilities expanded into the field of nuclear weapons.[42]

Much of the information about the North Korean program comes from Russian specialists with access to old Soviet intelligence. According to one account, Kim Il-Sung, the long-time North Korean leader who died in 1994, was impressed by how quickly the United States was able to terminate the war with Japan with two atomic bombs in contrast to his guerrillas running from the invincible Japanese army for fifteen years. Then, in the 1950s, his agents delivered documents indicating that the Truman Administration had considered using nuclear weapons against North Korea. In the 1960s, he was shocked by the Soviet Union's "abandonment" of Cuba, suggesting a need for greater military self-sufficiency. Finally, when the Americans deployed tactical nuclear weapons in South Korea, he saw little alternative but to respond with a North Korean capability.[43] Thus, Kim Il-Sung saw nuclear weapons as a "strategic equalizer," with targets in Japan, where U.S. interests could also be severely damaged. According to North Korean diplomats, he wanted only four nuclear devices, one for each of the principal islands of Japan.[44]

North Korea has not tested a device, but the infrastructure for producing nuclear material and components for use in weapons is substantial. Estimates as to the number of weapons they could now build, using plutonium produced in a small reactor, range from one to ten, although the actual number is probably one or two. Whether a device would be one year or five years into the future, the fear has been that if the program were to continue unchecked, the first device would be followed by many more.[45]

For a few years, North Korea, as a signatory to the Nuclear Nonproliferation Treaty, allowed international inspectors to visit some nuclear facilities. The inspectors consistently raised unanswered questions concerning the country's intentions related to weapons. Thus, when the North Koreans terminated the inspections in the 1990s rather than continue to be subjected to international acrimony, the United States aggressively sought alternative routes that would constrain weapons development.

In October 1994, after lengthy negotiations, the United States and North Korea signed a four-page Agreed Framework outlining some major horsetrading over key nuclear issues. North Korea pledged to freeze its nuclear program and comply with international requirements concerning safeguarding nuclear material and permitting inspections. It would also eventually dismantle its graphite-moderated reactors, which are capable of producing plutonium that could be used in weapons. In return, the United States pledged to arrange for the installation in North Korea of two large atomic power reactors (1,000 megawatts each) that use low-enriched uranium. They would neither use nor produce weapons-grade materials. Finally, both sides agreed to pursue technical issues concerning the storage and disposition of spent nuclear fuel—with potential weapons applications—currently in North Korea.

South Korea, a skeptical observer of this political process, has questioned the advisability of some of the provisions of the Agreed Framework, particularly those relating to financial assistance to its longtime enemy. With South Korea now being bailed out of its economic difficulties by the international community, the ability of the country to make any financial contribution to the reactor deal in the north—as had been anticipated—seems to have vanished, at least temporarily. Fortunately, the Agreed Framework has a flexible timetable which may salvage the agreement in the long run.[46]

The revelations about North Korean nuclear capabilities have injected major doses of instability into East Asian political relations. Given the entrenched mistrust between North Korea and its neighbors, further steps by that country to expand nuclear activities of any kind or to block implementation of the Agreed Framework will surely be viewed by its neighbors as attempts at political blackmail. At the same time, given the near-famine conditions in North Korea so vividly portrayed by television clips of children on the brink of starvation, additional investments in nuclear weapons by the government seems far less likely than in the past still this possibility cannot be totally dismissed.

• *The Israeli Nuclear Arsenal:* The best known *secret* among nuclear weapons analysts is the existence of a sizable, but unacknowledged, arsenal of nuclear weapons in Israel. Beginning in the 1950s, using a French-supplied nuclear reactor and an underground plutonium separation plant, Israeli specialists have been producing substantial quantities of plutonium, with some reports indicating that as many as 100 weapons are now held in their stockpile. Given the destructive power of a single weapon, this is a potent arsenal by any standards.[47]

Whether Israel would use such weapons is of course an unanswerable question. It is difficult to imagine a scenario whereby Israel would improve its position by using nuclear weapons against any of its traditional adversaries in the region. The reaction from all sides would undoubtedly be violent. The paradox is that the international community seems to accept Israeli weapons as a deterrent against Saddam Hussein and other foes. The actual use of Israeli, or indeed American, British, or French nuclear weapons, seems unacceptable.

- *Indian–Pakistani Nuclear Testing:* The announcements by the Indian and Pakistani governments in May 1998 that they had tested five and six nuclear warheads, respectively, have served as the most violent wake up call concerning the dangers of the proliferation of nuclear weapons materials and technologies. While their possession of nuclear capabilities has been well known for a number of years, their decisions to demonstrate a readiness to use nuclear weapons to protect vital national interests—and in Pakistan's case, a perception of their very survival—have added grave new dangers to political instability in South Asia. Achieving political settlements over the status of Kashmir and numerous other flashpoints in Indian–Pakistani relations must now not only cope with decades of mistrust and ethnic hatred but overcome a new type of mortal fear.

The initial international response to the tests has been a combination of political outrage and economic sanctions. The second phase of the reaction will undoubtedly be an attempt to persuade the two countries not to prepare nuclear weapons for use if they have not already done so. Then international pressure must be applied on the countries to sign the Comprehensive Test Ban Treaty, to join in international efforts to develop an international agreement to cap the production of any more uranium or plutonium, and to ensure that they do not transfer their weapons know-how or material to other countries of concern. Each of these steps will help in a modest way to ensure better control of nuclear capabilities not only in South Asia but in other regions of the world—such as the Middle East. Indeed, should India and Pakistan be allowed to run free and unleash a nuclear arms race, the prospects for controlling nuclear proliferation anywhere in the world will be very dim indeed.

While such short-term steps are readily identifiable, the longer term policy is much more difficult to define, let alone implement. What are the implications of the five traditional nuclear powers (United States, Russia, United Kingdom, France and China) willingly or unwillingly admitting two more countries to their club? In any event, India and Pakistan should not be allowed to achieve great power status through nuclear adventurism.

And what is the link to terrorism? At present it does not seem direct. But in time, a weakening of the international system to control nuclear technologies means that such technologies become more available to states or groups ready to use any weaponry to achieve their purposes with or without declarations of war. Also, as we will see, there have repeatedly been indications that drug-related money helped pay for the Pakistani nuclear program; and the step between drugs and terrorism is becoming shorter all the time.[48]

• *Reversing Course in South Africa:* In a unique move, South Africa did some classification-hopping by renouncing the nuclear weapons it had already developed. In 1993, the de Klerk government acknowledged South Africa had developed nuclear weapons. de Klerk emphasized that the weapons had been destroyed and the weapons program had been abandoned several years earlier. But, why would South Africa want to build seven nuclear weapons in the first place? According to South African participants in the program, South Africa planned to use the weapons as a means of intimidation to counter threats against the country. In particular, the South Africans were concerned that the former Soviet Union would attempt to expand its influence onto nearby territory. Should such a scenario unfold, South Africa planned to inform Western allies of the secret weapons and request Western aid in resisting impending aggression. If the allies were slow to respond, the South Africans would publicly announce their nuclear weapons capability, thereby thwarting Soviet adventurism directed against the country, so the government reasoned.

In one sense, the strategy was seen as a political bluff to force the United States or other Western powers into coming to South Africa's assistance. At the same time, South Africa could demonstrate that despite its isolation from much of the world due to the apartheid issue, it could conquer the most difficult technical problems on its own.[49]

- *Atoms for Peace, Even in Vietnam:* Vietnam is a non-nuclear state, presumably far removed from thoughts of nuclear weapons or nuclear power. Yet over the years, many poor countries have successfully used radioactive isotopes in investigating the properties of their agricultural crops and in diagnosing and treating hospital patients. I have witnessed firsthand a dozen developing countries flirting with attractive nuclear research ventures.

In the 1950s, Israel, India, and Pakistan were among the first to adopt programs promoted by the United States called Atoms for Peace, the concept being to provide certain countries with a small research reactor that could produce the radioactive isotopes useful in agriculture and medicine. Unfortunately, a few countries (including those already mentioned) subsequently chose to expand the peaceful atoms into military applications.

As for Vietnam, the U.S. Agency for International Development (USAID) dispatched me to Dalat, Vietnam, in 1972 to begin developing a nuclear program based on a small research reactor installed in the early 1960s as part of their Atoms for Peace Program. For ten years, the reactor was largely dormant. Even though the reactor had been under hostile fire during the Tet offensive in 1968, four years later State Department and USAID officials thought a program at the facility would symbolize U.S. confidence that the war was going in the right direction.

From the moment my heavily armed escorts and I boarded a CIA-chartered Air America cargo plane in Saigon, I was convinced that the whole idea of nuclear research in a war zone was a mistake. Upon landing at the airport in Dalat, I was greeted by an escort convoy featuring three U.S. army jeeps, each brandishing a 50-millimeter machine gun. Fighting was under way fifteen miles from the city and, therefore, the sightseeing schedule had to be cancelled.

At the nuclear reactor facility, two American-trained physicists spent three hours detailing their elaborate plans for a research program to develop isotopes for addressing agricultural and medical problems that would be facing postwar Vietnam. An early investment in research would reap substantial benefits when the country was free of war, they vigorously argued. All they needed was some additional equipment and a few more hands. For them the shooting in the distance was little more than background noise; they were prepared to tolerate such distractions. For me, the threat of radioactive debris being scattered from a mortar shell landing on the reactor overshadowed even the most optimistic predictions of how nuclear scientists could stimulate agricultural productivity. I thanked them for their presentation, knowing that their plans would never come to pass.

After a quick lunch in a secure bomb shelter in the city, I was hustled back to the Air America plane for the return flight to Saigon. I declined the pilot's thoughtful invitation to divert a few miles from the flight line to witness firsthand the jungle skirmishes nearby. Back in Saigon, I abandoned thoughts about the reactor in Dalat and turned my attention to another assignment that was also of questionable sustainability, namely, strengthening the ties between the local engineering college and the University of Missouri in Rolla.[50]

Once in Washington, I made a clear recommendation: get the nuclear material out of Dalat. I was not concerned about the theft of reactor fuel or of a small amount of loose plutonium. Eighty grams of plutonium would make only a minuscule contribution to a weapons program, I thought. Rather, I was focused on the radiation contamination of a large area should a shell destroy the reactor. A few weeks later, I received information from Saigon that the material had been moved to safe keeping—a report that turned out to be far from reliable.

Only in 1997—twenty-five years later—did an intrepid American reporter break the story of what had happened to the material. It was *not* removed from Dalat in 1972. Three years later, the U.S. governmented dispatched two experts to retrieve the fuel and the plutonium, lest the North Vietnamese find the material and embarrass the United States. The details of their entry into Vietnam have not been revealed, but the two engineers received awards for valor in wartime when they returned with nuclear material from what was considered a highly successful mission. However, it turned out that they mistakenly brought back *polonium*, a radioactive material far less dangerous than the plutonium they had left behind. Only in 1996 was the gaffe discovered by senior officials when they were reviewing old records being declassified, and diplomatic efforts were then launched to trace the lost plutonium. The latest investigation was still under way as of the end of 1997.[51]

The Vietnam caper illustrates the longstanding interest of countries at all levels of economic development in participating in nuclear activities, although in Vietnam's case, it seemed a sincere effort at solving agricultural problems. Some countries such as Argentina and Brazil have considered parlaying their nuclear scientific capabilities into weapons programs, but fortunately have decided to push such ambitions aside. Others, in addition to those already mentioned, and particularly Iran, are believed to covet the nuclear weapon and must be carefully watched.

Possession of a nuclear program has always had a dimension of prestige. Nations want to be seated at the international table when the future of the world is discussed. Institutions that possess special types of nuclear material gain recognition at home and abroad—with certain rights and obligations. Our latest concern, however, puts a fresh perspective on the table. Prestige must take second chair to the very real threat that hangs over us: the specter of a new brand of terrorist intent on brandishing power with black market megabombs.

Until now we have had on our side the fact that nuclear weapon design and construction are complicated and expensive tasks, requiring vast teams of specialists and sophisticated facilities and equipment. Should a terrorist group undertake such a project, their efforts would involve complicity of a government and would quickly be detected by other governments early in the process. We would then pull out our traditional menu of diplomatic choices, backed by military force, for dealing with radical regimes. So went the logic of the past.

Now fundamental assumptions about the essential role of a government in the support of any nuclear weapons activity are being challenged. Yes, constructing a nuclear device remains a complicated task, and such an undertaking requires substantial financial and engineering resources. A nuclear bomb cannot be built in someone's garage. However, recent developments ease the process: (1) basic design approaches are known worldwide (2) an increasing number of scientists and engineers all over the globe are willing and able to take on the task; and (3) specialized equipment needed to support the task is becoming more widely available.

Rogue regimes with an array of technical resources remain the most immediate concern. We can no longer assume that the broadly based technological approaches of the past requiring hundreds of skilled specialists working in sophisticated facilities—a Manhattan Project if you will—will be the only avenue in the future. We surely cannot assume that some or all of the ingredients for weapons will not be purchased or stolen. Nor can we assume that such threatening efforts will be limited to programs carried out under the patronage of rogue governments and on territory controlled by these governments. Given these realities, how can we respond?

- World-wide controls on exports of key materials and equipment to countries and groups of concern are central to limiting capabilities to produce nuclear weapons. Severe economic sanctions should be applied against those exporter countries who fail to enforce such controls.
- Improved intelligence and international inspections under IAEA auspices are crucial to upgrading our capacity to uncover illicit activities.
- U.S. assistance efforts to reduce and consolidate stockpiles of nuclear warheads and of *HEU* and plutonium of other countries, particularly the states of the former Soviet Union, should be expanded. Also, increased cooperative efforts to upgrade systems that ensure the security of such items should be undertaken immediately.
- The same vigor unleashed to apprehend murderers of law enforcement officers should be applied to the international pursuit and prosecution of thieves seeking nuclear material.

Stronger international conventions and additional extradition treaties must be put in place for interdicting nuclear smuggling and for prosecution of nuclear terrorists.

Meanwhile, we buttress our defenses at home. One domestic initiative in place since the 1970s is the Nuclear Emergency Search Team (NEST) of the Department of Energy. They are charged with responding to the threat of nuclear terrorism or other nuclear incidents in the United States. Two personal vignettes demonstrate how it operates:

"Do not let that plane take off without my personal permission!" The administrator of the Environmental Protection Agency (EPA) in Washington, D.C. was issuing her telephone instructions to me in my director's office at EPA's laboratory in Las Vegas, Nevada, in 1983. Fifty of the laboratory's radiation monitoring specialists stood by at nearby McCarran Airport. They were poised to fly to the yet-to-be-determined site of an imminent crash landing of a wayward Russian satellite that relied on a nuclear power source.

The predicted crash site was in California where the debris from the nuclear device, expected to disintegrate upon re-entry into the atmosphere, was likely to be scattered over a five hundred mile trace. The administrator had taken the unusual step of giving me her confidential home telephone number to be sure she had a chance to inform the governor of California and the media before her environmental shock troops arrived. Battered daily by the press since her arrival at the agency for EPA's slowdown of regulatory actions to reduce air and water pollution, she planned to respond to this impending disaster in a way that would demonstrate her ability to act quickly and decisively during a major emergency.

The EPA contingent was part of NEST. Four years earlier, EPA specialists on behalf of NEST had tracked down the debris of another faulty Soviet nuclear satellite that had crashed in an uninhabited region of northern Canada. This time, however, the concerns were far more real. The extent to which radioactive fragments would cause a threat to the population was not predictable, but press speculation would certainly cause turmoil throughout California.

In the end, the satellite belied predictions and quietly splashed down in the Indian Ocean. The EPA specialists returned to their more routine jobs of monitoring conditions at the nuclear weapons testing area in southern Nevada.

One year later, NEST duties again called in laboratory specialists to help prevent nuclear calamity. Security measures for the Los Angeles Olympic Games included searching for nuclear devices that might be smuggled into the area by terrorists. The radiation specialists were to respond to tips received by the counter-terrorism control center in downtown Los Angeles. Again the shock troops were standing by. Fortunately, for two weeks there were no indications of nuclear terrorism.

When NEST was established their theory was an extortionist would allow several days for negotiation. During that time, NEST's detection specialists would be able to locate and methodically disable the nuclear device. Then in 1980, a very sophisticated non-nuclear explosive device was detonated in Harvey's Casino in State Line, Nevada, killing several security personnel. At the time, local authorities were attempting to disarm the device. This incident spurred NEST to develop better equipment and techniques to quickly render harmless high-technology devices of all types.

Now the emphasis is rightly placed on terrorism-without-notice threats rather than more prolonged extortion scenarios. A frequent prediction is that terrorists will mix radioactive material with high explosives, and nuclear contamination will be scattered over many square miles. This scenario will not cause nearly the amount of damage as detonation of a nuclear bomb, but depending on the violence of the explosion and the density of the nearby population, the radioactivity could claim many victims.

Whether searching for radioactive debris, providing security for sporting events, or disarming explosive devices, NEST will play a key role in bringing technical assets to any counter-terrorism strategy.[52]

Many suggestions have been made as to additional steps that should be taken to forestall the day nuclear terrorism erupts in the United States and to reduce the likelihood that terrorist efforts will be successful. Among the suggestions are: (1) provide the Customs Service with more sophisticated monitoring devices for border surveillance; (2) improve "fingerprinting" techniques to reveal the exact composition of nuclear material for determining the source of seized contraband material and (3) use retired weapons designers in on-call strike forces that respond to terrorist threats.[53]

Some nuclear weapons specialists argue that the day has already arrived when nuclear weapons are available to groups that do not have broadly based technological capabilities. They cite reports (which I call unsubstantiated rumors) that the Islamic Jihad tried to purchase a nuclear device from a Russian weapons laboratory and that the Palestine Liberation Organization (PLO) has one or two nuclear weapons buried in the desert. Such allegations feed the intelligence frenzy surrounding the future of nuclear terrorism.[54] It is certainly true that the Aum Shinrikyo explored the possibility of developing a nuclear capability, sending exploration teams to Australia in search of uranium.[55]

Thus far we have been lucky in keeping nuclear weaponry from being used by rogue regimes or independent terrorist groups. But as violence and frustrations mount throughout the world, we can no longer afford to rely on luck.

Some Americans who fear a nuclear armageddon plead for the complete elimination of nuclear weapons from the earth. In one of the most famous—and controversial—presidential statements of all times, President Reagan long ago called for ridding the world of nuclear weapons. Mikhail Gorbachev joined him in the appeal when they met in Iceland in 1985. Reagan's advisers, taken by surprise, argued that diplomacy doesn't work that way. They quickly buried his statement under an icy avalanche of denials, presented the public with frosty explanations of poor staff work, and bellowed chilly reminders of the evils of communism that lurked behind every Gorbachev pronouncement. A polite statement by Secretary of State George Shultz highlighted the merits of the president's position but noted that partial and not complete denuclearization was the urgent imperative.[56]

Today, a number of hard-bitten generals who led our armed forces during the Reagan years, in a new alliance with misty-eyed idealists, are also calling for movement toward a zero level of nuclear armaments.[57] Using the best of military jargon in addressing the likelihood of nuclear war, they argue, "The most commonly postulated nuclear threats are not susceptible to deterrence or are simply not credible."[58]

The generals and idealists believe that renegade states or terrorist groups armed with nuclear capability will not be intimidated by our ability to obliterate them after they have destroyed a large city.

Could holding on to our nuclear weapons because other nuclear wannabes are obtaining nuclear capability eventually lead to the end of the world? Do the risks of nuclear accidents or the acquisition by an irresponsible party of nuclear weapons far outweigh the value of having such weapons? Does reason demand these weapons be discarded?

Most political and military leaders argue persuasively for a *less*-nuclear rather than a *non*-nuclear world. Nuclear technology cannot be unlearned. In their view, some countries and some terrorist groups will always be seeking to acquire destructive nuclear devices. If such terrorists alone have nuclear devices, they can hold the rest of the world hostage to their political demands. Thus U.S. leaders will surely reject efforts to reduce our nuclear stockpile to zero. We must be prepared—if only for purposes of self-defense—to share our precarious world with others who also have nuclear weapons.

That said, we can have no higher immediate priority than keeping nuclear weapons and the capability to build nuclear weapons out of the hands of hostile people. We cannot forget that in some cases a reliable state today may become a rogue state tomorrow. Also, a rogue state may be but a telephone call or an e-mail message away from a terrorist group willing to consider a mutually beneficial joint venture.

As we are quick to brand as being terrorist those rogue nations seeking nuclear capability, we ourselves should be prepared to wear the terrorist name badge should we threaten to use nuclear weapons irresponsibly. In particular, when responding to chemical, biological, or conventional attacks, even if they are very serious, we should forget the nuclear option. The stakes of a nuclear holocaust are even higher and we cannot tolerate nuclear adventurism—in word or deed. To do otherwise would simply legitimatize a free-for-all in unleashing indiscriminate use of the most lethal technology ever devised.

CHAPTER 3

Chemoterrorism:
Poisoning the Air, Water,
and Food Supply

I'm frightened by chemical weapons—the most vicious
weapons of all. You can't hear or smell them. First thing
you know people are dropping dead. We need a great
outcry against them.
Billy Graham, 1996

Japanese Cult Said To Have Planned Nerve Gas
Attacks in U.S.
The New York Times, 1997

We were totally unprepared to respond to the threat of a
poison gas attack at Disneyland in 1995.
Senior official, Department of Health
and Human Services, 1997

*A*ccompanied by a chorus of shrieking sirens, police cars and
ambulances descend on the Woodley Park Metro Station in
Washington, D.C., responding to reports of what the police have
been told is a subway train crash. A pedestrian stretched out on the
sidewalk vomits a growing mass of bubbling froth. Several others
nearby are having convulsions. Another dozen or so shakily clutch
the handrails of the escalator. The emergency responders themselves are becoming ill, and the police grab their radios to call for
backup. One of the paramedics murmurs in disbelief, "These
symptoms almost look like the result of a nerve gas attack."

Eventually, the Fire Department's HazMat (hazardous materials) unit arrives. Captain Thomas Johnson, a towering man on any day, steps from the HazMat truck in full Level A gear—a sealed space suit and mask. The sight of him in this ominous garb sends the gathering of spectators scrambling for cover. The unprotected police make tracks, though they do not know where to flee.

The experienced Johnson has already mastered his new handheld chemical detector. The device indicates the presence of a nerve agent—probably sarin, he guesses—and he calls for a full-scale chemical-biological response. Twenty minutes pass while a dozen other police and fire officials retrieve their HazMat suits from the truck and don them. Within an hour, 100 officials are surrounding the scene at what they hope are safe distances, evacuating many blocks along Connecticut Avenue.

Meanwhile, 911 is becoming flooded with reports.. One panic-stricken caller claims to have seen suspected terrorists spraying gas from a boat on the Tidal Basin, while others report people collapsing on the mall. It finally becomes clear sarin was released within the Metro system, and the Metro stations along the Red Line must be promptly evacuated. The death toll mounts to several hundred, with thousands illnesses already reported.

- Where is the Metropolitan Medical Strike Team, the model for the country? It is still not activated due to lack of training.
- Where are the support units from nearby military facilities supposedly available for this type of incident? Their role in such civilian disasters is still unresolved.
- Where are the decontamination showers? The hospitals have long waiting lines for proper shower stalls. There are not enough stalls to handle such a mass influx of emergency victims.
- Who is in charge? The White House is asking, but no one knows the answer.[1]

This well-researched scenario was published in a Washington tabloid in March 1997. According to the author eight hundred deaths would occur in the first twenty-four hours, with eleven thousand additional illnesses. Downtown areas would become off-limits decontamination zones for unknown periods of time, while they were being examined and detoxified. The Metro system itself would be closed for an even longer period.[2]

Perhaps in the face of a real incident, the city's responders, despite their shortfalls in preparations, would perform at a higher level than expected. But then again perhaps terrorists, taking adequate time to prepare, would have sought even greater damage from their plot, including additional releases or explosions to disrupt the emergency measures.

Precisely one week after publication of the hypothetical sarin attack, testimony in a Tokyo courtroom revealed the Aum Shinrikyo religious cult had planned to unleash sarin in the United States.[3]

Five weeks later the Marine Corps' Chemical/Biological Incident Response Force staged and responded to a mock chemical attack on Capitol Hill. This is the only team of its kind in the United States. Its commander boasts they are the ones who put on the gear and move down range to the hot zone. In his words, "we take contaminated victims and turn them into decontaminated patients."[4]

Even though the death and devastation from the sarin attack in the Tokyo subway system have been recounted many times, this watershed event heralding the emergence of a new age of high tech terrorism is still shocking to behold. Most important, it teaches us many lessons.

Five million passengers use Tokyo's subway system every day. And every day trains packed with commuters converge on the Kasumigaseki Station in the heart of the city near several government ministries. During the morning rush hour on March 20, 1995, five members of the Aum Shinrikyo—all well-trained Japanese scientists or engineers—carried in a total of eleven small plastic bags, laden with a chemical solution containing deadly sarin, onto five trains racing for the station from different directions. Each chemoterrorist placed his set of bags under a seat or on a baggage rack and, as his train approached the station, punctured each bag with a sharpened umbrella tip. They then quickly left the trains and returned to a hideout where they received injections of a sarin antidote as they reported to their leader. The final toll of their work was twelve deaths and 5,500 injuries.[5]

This incident has been analyzed and reanalyzed in the United States and around the world.[6] It has provoked warnings from terrorism experts in many countries that we have definitely crossed a threshold to *a new frontier of superterrorism for the year 2000 and beyond*. Some describe the incident as the realization of the nightmare scenario experts had quietly talked about for years.[7]

Chemoterrorism, according to the former director of the CIA, is the most likely form of high-tech attack to erupt in the near term in the United States.[8] During congressional hearings in the wake of the Tokyo incident an array of senators posed the following questions to the CIA and other government agencies aimed at the heart of America's readiness to deal with this new type of attack.

- How was the doomsday cult able to recruit university-trained scientists in Japan and elsewhere, and what are the implications for Western industrialized countries?
- What did U.S. law enforcement and intelligence agencies know about the intentions and capabilities of this group before the Tokyo incident?
- How could a religious group with avowed intentions to provoke attacks not only in Japan but also in the United States accumulate such technology and weaponry in a relatively short period of time without raising alarms in U.S. intelligence and law enforcement agencies?
- Was this cult linked to or supported by other groups, whether political activists or criminals?
- Could such an event happen in the United States?
- If so, are our intelligence, law enforcement, and public health agencies prepared?[9]

For two years, the U.S. Senate searched for answers to these questions. Staff members and witnesses prepared extensive documentation and testified on these issues and other aspects of the incident. Intelligence agencies may now have 20/20 hindsight on the problem, but nagging questions linger.

How could it possibly be that this cult, with more than sixty thousand members in Japan and other countries and possessing assets of more than $1 billion, was not of interest to American intelligence agencies prior to the incident? Eight months before the

Tokyo incident this cult had launched a "dry run" in Matsumoto, Japan and killed seven people with a sarin release. The cult's involvement in that incident was well publicized in the Japanese press several months before the Tokyo attack, and the police had even begun to take initial steps to disrupt the cult's future terrorist activities. (See Figure 5.)

Of course, the Aum Shinrikyo attack was, in the first instance, a Japanese problem. But given the Aum Shinrikyo's activities around the world, including activities in the United States, the cult certainly should have been on the radar screens of American intelligence agencies prior to the Tokyo disaster. But it wasn't— even after the Matsumoto attack, which was the biggest news story in Japan for weeks. It was the first non-military use of nerve gas in the world. And sarin was released in a densely populated country where tens of thousands of Americans live and work, including units of the U.S. armed forces.[10]

The Aum Shinrikyo had established an office in New York City in 1988, with links to sympathizers in Los Angeles. In addition to distributing literature about revolutionary religious views (e.g., *The Doom's Day* and *The Secret Method To Develop Your Super-human Power*), the office procured both high technology and low technology items with possible military applications. For example, their order of two hundred Israeli gas masks from an American store specializing in surplus military items, while not illegal, should have raised a red flag. Apparently the sale was never brought to the attention of law enforcement agencies. Meanwhile, the cult's monthly paper, *Vajrayana Sacca*, was publishing in Japan a series of anti-Japanese and anti-American articles with one report in January 1995 proclaiming:

> Clinton will be without doubt a one-term president. At best, he will not be re-elected. At worst, it would not be strange if he were assassinated, making it appear like an accident.[11]

Another puzzling aspect is the disinterest of American intelligence officials in the cult's ties to Russia. There the Aum Shinrikyo claims a membership of at least thirty thousand. Beginning in 1992, it established contacts with several leading Russian political figures. The cult's acquisitions from Russia included a helicopter, presumably to be used for spraying chemical or biological

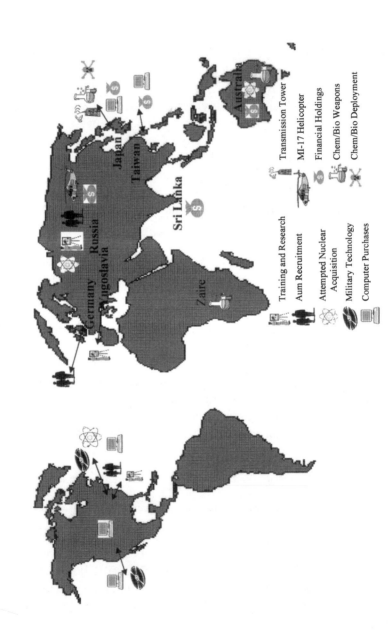

5. Aum Shinrikyo Global Activities

agents, as well as pistols and other light weaponry. Also, Aum Shinrikyo members from Japan traveled to Russia for training in military tactics.[12]

As to the cult's future status, the Japanese government apparently is prepared to let the cult continue its legally recognized existence, despite its avowed intent to destroy the government. The Japanese police claim they are determined not to be caught napping a second time.[13] They should also be aware that in Russia, a core of believers may be continuing to conspire as to how to make money by selling technologies and services to their suspect colleagues in Japan.[14] In the United States, a few years after the incident, a host of federal agencies have responded to the Senate probe and other inquiries by reassuring Americans that Washington is raising its guard by establishing task forces, strike forces, and response teams. They surely have elevated chemoterrorism to the top of the nation's security agenda, but they make no promises about preventing any further incidents. Intelligence agencies should no longer turn a blind eye to the cult's future practices or to the activities of other well-financed and similarly motivated zealots elsewhere in the world.

Dealing with groups whose religious and destructive political agendas coincide is never easy, particularly in Japan, which has a high level of tolerance of religious fanaticism. Violent organizations like Aum Shinrikyo, with chemoterrorism on their minds and millions of dollars in their bank accounts, have not been common in the past. The challenge to U.S. law enforcement agencies is to work with other governments in unearthing illicit acts hidden behind smokescreens of legitimacy.

Prior to the Aum Shinrikyo's bringing chemical agents into the spotlight, almost all perpetrated incidents involving the malevolent use of chemical agents were by individual terrorists and small terrorist groups using poisons on a minor scale. Many of the reported events since 1970 are noted in Table 4.

In the not too distant future, chemical incidents may well escalate into attacks of much broader reach in the United States and abroad. Terrorists may believe because some countries will react to shootings and bombings as almost expected incidents, they will need more spectacular events to send strong political messages. At the same time, as demonstrated in Tokyo, some violent criminals

TABLE 4.   Examples of Use of Chemical Poisons

Incidents in Israel (1970–90)
- Palestinian terrorists, supported by Iraqis, poisoned exported Israeli citrus fruit with mercury.
- A carbamate pesticide was found in the coffee in a military dining hall.
- In a Haifa market, Gerber baby food was found contaminated with the pesticide methomyl.
- Rodent poison was placed in drugs in Jaffa.
- A Palestinian worker contaminated food articles with the pesticide parathion in a Jerusalem minimarket.
- Contamination of bottled drinks was reported in a factory near Tel Aviv.
- Turkeys at a farm in Shalva consumed water that had been deliberately contaminated.

Source: Dany Shoham, Chemical/Biological Terrorism: An Old but Growing Threat in the Middle East and Elsewhere, *Politics and the Life Sciences*, September 1996, pp. 218–19.

Other Incidents
- The city government of Phoenix received a threat that its water supply would be contaminated if extortion payments were not made; no payments were made (1978).
- Cyanide-laced Tylenol capsules were discovered in the United States (1982).
- A guerrilla group reported that tea from Sri Lanka was poisoned with potassium cyanide; no contaminated tea was found (1982).
- In their protests against animal experiments, the Animal Liberation Front contacted the Mars Company to claim they had spiked the company's candy bars with rat poison; no poison was found, but the scare cost Mars $4.5 million (1984).
- Fruit from Chile on its way to Japan and the United States was reported poisoned; only two cyanide-laced grapes were found, but the scare cost Chile $333 million (1989).

Source: Roger Medd and Frank Goldstein, International Terrorism on the Eve of a New Millennium, *Studies in Conflict and Terrorism*, July–September 1997, p. 284.

are less concerned about deaths of innocent people when they target broad population centers. Dangerous chemical technologies are becoming increasingly available to all.[15]

In 1985, the FBI uncovered the most significant, documented case of an American-based terrorist group planning to use a chemical agent to date. The Covenant, Sword, and Arm of the Law was a white-supremacist, anti-Semitic group with a housing compound in the Ozark Mountains of northern Arkansas. During a search of the compound, federal agents recovered drums containing thirty-five gallons of cyanide. The group apparently planned to dump these into the water supply of either Washington, D.C., or New York City. Subsequent studies done to determine the effect of the cyanide in a huge volume of water indicated that the dilution factor would be so great that there would have been no effect.[16] While cyanide would be readily diluted to impotency if dispersed in reservoirs, water supply lines have many vulnerabilities in more immediate proximity to consumers. Such a possibility should clearly be a concern of public officials.

In the United States, as in most industrialized countries, *accidental* chemical poisoning incidents have become a way of life, with daily reports of misuse of cleaning agents, pesticides, and pharmaceuticals, not to mention illegal disposal practices. A recent egregious example occurred in Mississippi where a pest exterminator sprayed methyl parathion in more than four hundred buildings. This neurotoxin does, indeed, kill everything that slithers, crawls, or even walks—including human beings, when administered in huge doses. As the exterminator sat in jail, the cleanup bill for the concoction, dubbed "cotton poison," was assumed by the U.S. Environmental Protection Agency and eventually reached $40 million.[17] Despite plenty of labeling laws to help reduce the likelihood of such accidents, scofflaws will always exist. We must keep in mind that warning labels hardly equate with prevention.

Until the Tokyo incident, Americans were relatively secure in the belief that *deliberate* poisoning was only an occasional occurrence perpetrated by a single, severely disturbed individual. While we would like to continue to think that such misapplications of poisons will be relatively rare, the sarin gas attack puts us on notice of the fragility of that notion.

The Director of Emergency Services of New York City, accustomed to responding to accidents involving chemical toxins, heard the wake-up call from Tokyo loud and clear. He has warned it would be simple for even an untrained terrorist to dump parathion into the heating or air conditioning system of a high-rise building. According to the director, the poison could kill just about everybody in a high-rise building that houses six thousand residents, and that's just *one* well-conceived attack.[18]

From a different perspective, what could the Israeli intelligence service, Mossad, possibly have been thinking when they attempted to kill the Hamas leader Khaled Meshal in Jordan with a lethal nerve agent? In fall 1997, five witnesses saw a Mossad agent inject the toxin into the skin behind Meshal's ear as he walked down the street. The agent was captured as he fled the scene. Within a few hours, Meshal had uncontrollable vomiting and went into respiratory arrest. He lay perilously close to death in a military hospital. Israel was then compelled to meet Jordanian and American demands to supply the antidote—also carried by Mossad agents—in an extraordinary admission of the botched assassination attempt. Meshal survived.

While Israel may have had its own reasons to retaliate for Hamas bombings in Tel Aviv, why use a nerve agent rather than the tried and true method of a bullet? And why on Jordanian soil? Was the delayed effect supposed to provide time for the perpetrators to escape undetected? Perhaps Mossad needed a cover story, explaining that the Hamas leader died of a strange disease. In short, if Mossad can't effectively handle poison agents, one can only surmise as to the chaos that will ensue if less experienced organizations try their hands at chemical attacks. Antidotes have yet to be developed for many chemical agents.[19]

The likelihood of other types of state-sponsored chemoterrorism was driven home to me on a more personal level in 1992. I met a senior Libyan chemical engineer working in Moscow. He lived across the hall from us in our apartment building. Ostensibly, he had been studying pesticide production in Russia for four years, with a government job awaiting him upon return to Libya. Beyond these bare facts, he was reluctant to describe his actual work. With Russian industry in shambles at the time, Moscow seemed to me like a strange choice for studying at such length the workings of the chemical industry, although the Russians certainly knew how to produce lethal pesticides.[20]

In considering state-sponsored chemoterrorism, the experience of military organizations, both in using chemical agents and in countering chemical threats on the battlefield, will take on greater importance than ever before. Over the centuries, some military commanders have hesitated to unleash chemical agents on adversaries—sometimes respecting a higher moral standard but in recent times, more often fearful of retaliation. Indeed, victory by any means has been a tenet of many armies. Twenty-five hundred years ago, Athenian troops poisoned water supplies, and Sparta allies seized a fort after directing smoke from coal, sulfur, and pitch through a hollow beam into the fort. Five hundred years ago, Venice also poisoned drinking wells and destroyed crops and animals of its enemies with lethal chemicals. Napoleon reportedly considered dipping bayonets into cyanide. During the American Civil War, some Union generals advocated loading hydrochloric acid and chlorine into projectiles.[21]

The use of chemical weapons as a tool of modern warfare burst onto the battlefields of Belgium in 1915. The hissing sound from the German trenches was the release of the choking agent chlorine from six thousand cylinders directed toward the British and French trenches downwind from the release. The situation within the totally unprepared ranks of allied soldiers has been described as follows:

> Soldiers clutched their throats, their eyes staring out in terror and pain. Many collapsed in the bottom of their trenches and others clambered out and staggered to the rear in attempts to escape the deadly cloud. Those left in the trenches writhed with agony, their faces plum colored, while they coughed blood from their lungs.[22]

While fewer than a thousand deaths resulted, more than fifteen thousand horrified troops retreated in a state of panic.

In 1917, the Germans again surprised the allied forces by delivering mustard gas in artillery shells onto the battlefield. The gas was persistent. Not only did the air that the troops breathed become pathways of exposure but so did the objects they touched. Also, the latency period of several hours from the time of exposure to the agent to the manifestation of the gruesome symptoms to come meant there were no immediate clues as to who might develop blisters or suffer gastrointestinal and other effects.

To a limited extent, the allies responded by releasing chlorine and cyanide of their own. By 1918, hundreds of thousands of soldiers and civilians had been injured by chemical weapons, and up to 100,000 had died.[23]

The Italians used chemical weapons in Ethiopia in 1937, and the Egyptians employed them in Yemen in the 1960s. During the Vietnam War, American spraying of Agent Orange and other herbicides as jungle defoliants and American firing of riot control agents to quell demonstrations in American cities stirred controversy at home and abroad. In the 1970s the Libyans employed chemical agents in their war with Chad. Then in 1983, Iraq began using chemical agents (mustard, tabun, and sarin) against Iran on a more significant scale, and Iran retaliated. Also, according to many press reports, the Iraqis used mustard gas against the Kurds in the 1980s.[24]

Over *twenty* nations now have the capability to produce and use chemical weapons.[25] Unfortunately, some terrorists are positioned to take advantage of these advances—terrorists who may be military personnel, who may be former soldiers, or who may work hand-in-glove with armies of rogue states. As indicated in the Tokyo attack, deadly technologies have become available to

6. An Iraqi field strewn with scuds able to carry chemical warheads.

scientists throughout the world. Scientists can be quick studies in the art and craft of chemoterrorism even though they may not get all the details right the first time.

The most common chemical warfare agents developed by armies around the world can be divided into: choking agents (such as chlorine), blood agents (hydrogen cyanide), nerve agents (sarin), and blister agents (mustard). As suggested by each category's name, each type has different effects when it comes into contact with skin or when inhaled or ingested.

Other nonlethal chemical agents are also available but receive less attention by the public as tools of terrorists. Tear gas, vomiting agents (arsenic-based), and psychochemicals (hallucinogenic compounds), while not deadly, certainly can create the panic and pandemonium on which terrorists thrive.[26]

Each chemical agent has certain properties that determine how it can best be unleashed, and the recipes for terrorist attacks will vary accordingly. Hydrogen cyanide vaporizes quickly so it has limited outdoor use, whereas mustard gas persists so it can effectively contaminate a large outdoor area. Some gases can be delivered using explosives or sprayers. Others can be simply poured onto a surface. Volatile chemicals with even moderate toxicity released in unventilated spaces can cause great damage. Basic information on lethal agents has been available for many years in chemistry libraries throughout the world. A new development is the ready availability of information on successful and unsuccessful approaches to incorporating chemical agents into handy weapons—information now dispersed on the Internet. It may also be accessible from some amoral, underemployed former soldiers who seek to join a new wave of mercenaries.[27]

Given the direct involvement of armies in perfecting the use of chemical agents and in bracing themselves to cope with exposures of troops on the battlefield, these same armies must take the lead in developing responses to attacks at civilian locations. Though we would not encourage the use of soldiers to routinely patrol the streets of America, special military units can play critical roles in training civilian groups and in providing a logistical base of support in the event of incidents. (Their roles are particularly important in supporting response teams as discussed in Chapter 9.)

The United States and Russia have accumulated the world's largest arsenals of chemical warfare agents, covering all of the aforementioned agents to some degree. Now, the two countries have embarked on multibillion dollar efforts to destroy their stockpiles. Understandably, the American effort is moving forward at a far more rapid pace than the Russian program which, like most activities in that country, is financially strapped and barely off the ground. We are reasonably confident the American stockpiles of chemical agents are guarded by responsible security forces and that any method of destruction will be relatively safe for the populations surrounding these sites.

However, the present security of Russian stockpiles is suspect. The timetables for chemical destruction and for cleanups of contaminated facilities will likely be drawn out over decades. The Russians claim that the price tag for eliminating the stockpiles is over $5 billion, with initial steps measured in hundreds of millions of dollars per year. Fundamentally, the Russian financial resources required to take effective actions in the interim to ensure that toxic agents do not leak into the hands of terrorists or poison the environment simply do not exist.[28]

Reports by American specialists indicate that the security at the six or so sites in Russia where hundreds of tons of chemical agents are stored is reminiscent of security at American sites forty years ago—before the advent of tamper-proof locks, closed-circuit TV, and alarm systems sensitive to sounds and movements. Nevertheless, the Russian system of bars and padlocks, gates and guards, and sign-in/sign-out sheets would seem adequate were it not for the collapse of the economic and security structures within the country and the rampage of crime and corruption throughout the society.[29] Without modern physical security systems, the temptations for insiders to prepare their own orders of "carry-out" chemical weapons or for outsiders to penetrate the facilities seem too great. This situation appears to parallel conditions at some nuclear facilities.

While the Aum Shinrikyo may have prepared their own sarin cocktails, the recipes for premixed Russian versions, should they be available, would save such groups a great deal of time and effort and would deliver a far more lethal blow. And with fat bank accounts, the Aum Shinrikyo would not have been constrained in using high-level bribery to obtain such recipes.

The details of the Russian destruction techniques themselves are an entirely different matter. Members of the environmentally conscious political parties in Russia have seized on the destruction issue as a centerpiece of their sputtering emergence as political players. They demand persuasive data that there will be no ecological side effects if the chemicals are burned, detoxified with other chemicals, or (supposedly) rendered harmless through bioremediation technologies. For example, convincing environmental activists—the Greens—there will be no health or ecological risks in populated regions from gases during incineration of the agents, should this technology be adopted, would be a tough sell. The Greens' local allies in questioning any government proposal include many of the same workers at the chemical plants that originally produced sarin and other agents—workers who have suffered ill effects and now claim they should receive disability pensions due to their exposure for many years to low levels of these agents.

The physician responsible for establishing and enforcing worker exposure levels in the Volgograd plant where considerable chemical weapons production was carried out told me in late 1996 that she agreed with the claims of the workers. In the rush to produce weapons to fill their hefty quotas, enterprise directors rarely took time to enforce occupational health standards within the factories. Besides, people were hardly considered high-priority items in the Soviet industrial machine. She added that the Duma, the lower house of the Russian parliament, would surely support the workers and strongly oppose future acts that could have negative health consequences, regardless of the price tags of alternative approaches.[30]

Another major cache of chemical weapons has made the news in Iraq. Whether this is a *real* cache or a *virtual* supply of disassembled components that could be reconstructed within a few months is a topic of heated debate between UN inspectors (who have been on the scene since 1992) and Iraqi authorities. In late 1997 and early 1998, Saddam Hussein's reluctance to open seventy presidential sites set aside for his personal use, including one the size of Washington, D.C., casts doubts on Iraq's contention that it destroyed its chemical weapons at the end of the Persian Gulf War, thereby complying with international requirements to dismantle all weapons of mass destruction.[31] Indeed, his position triggered the possibility of yet another military confrontation with Western forces.

UNSCOM has reported that work leading to the production of the nerve agent VX began in 1985 or earlier, and, in 1990 alone, Iraq produced sixty-five tons of chlorine for use in the VX program. Hundreds of tons of two other chemicals needed for production of VX were produced: phosphorous pentasulphide and di-isopropylamine. Combined, these quantities of the three chemicals could support production of more than four hundred tons of VX—which translates into deaths for millions of potentially exposed victims, according to U.N. experts. While VX has captured the primary attention of the U.N. inspectors, other agents have also been developed.[32] (See Table 5.)

The Iraqi experience in developing the capacity to produce chemical weapons, under the guise of building a pesticides industry, is very instructive. Other rogue countries, such as North Korea and Libya, with genuine needs for pesticides and pharmaceuticals, may also be hiding nerve agents behind a smokescreen of legitimacy. They have learned from the Iraqi model that equipment can be obtained directly or indirectly from many foreign companies that are more interested in cash than worried about clandestine production of chemical weapons.

TABLE 5. Iraqi Chemical Agent Stockpiles

	Declared by Iraq	Potential Additional Agents Based on Unaccounted Precursors
VX	at least 4 tons	200 tons
Sarin	100 to 150 tons	200 tons
Mustard	500 to 600 tons	200 tons

Note: Estimates are in metric tons. Much has been destroyed. The data are derived from UNSCOM reports. Gaps in Iraqi disclosures suggest Baghdad is concealing agents and precursors. Baghdad has the capability to quickly resume production at sites that produce pharmaceuticals and pesticides.
Source: *Iraqi Weapons of Mass Destruction*, February 13, 1998. No attribution but released by CIA in February 1998.

The list of international firms that supplied the Iraqi Establishment for the Production of Pesticides has been described as a sheaf of pages out of a European telephone book. The companies provided building supplies, pumps, vessels, centrifuges, pipes, and chemicals. According to U.S. data, 207 firms from twenty-one countries (including some from the United States) in some way supported the growth of Iraq's chemical and biological programs.[33]

However, the Iraqi efforts were far from trouble-free. Reportedly, one-third of the chemical workers each year suffered occupational injuries while producing chemical weapons.[34] Some of the injuries were undoubtedly linked to development of prototypes of sarin-filled artillery shells, 122-millimeter rockets, and bombs. Also, accidents were purportedly commonplace in Iraq's program to flight test long-range missiles with chemical warheads, including one version filled with sarin. Presumably these personnel were subjected to some degree of military discipline, suggesting that the toll among more free-wheeling terrorists attempting to mount a chemical attack would be even higher.[35]

Of course, the existence during the Persian Gulf War of a variety of chemical warheads in Iraq is at the heart of the controversy over possible toxic exposures of American troops in the desert. Czech reports of samples collected during Desert Storm contain persuasive evidence of sarin gas releases from bombing runs. The levels of exposure of troops, however, to the chemical and the resulting health effects of the exposures, if any, will remain a controversial topic for years to come—though there are several studies attempting to determine just what may have happened. It is too difficult to reconstruct exposure levels and to separate medical ailments due to chemical warfare agents from those due to other factors.[36]

During the early 1990s, the international community forged a new international agreement to ban the production, acquisition, stockpiling, transfer, and use of chemical weapons. This agreement, the *Chemical Weapons Convention,* is one of the most ambitious treaties in the history of arms control. Whereas most treaties limit only the number or deployment of weapons, this agreement requires the outright elimination of an entire category of weapons. The parties to the Convention must destroy *all* chemical weapons and weapons production facilities which exist while foreswearing any intention to produce such weapons in the future. A new international organization, the Organization for the Prohibition of Chemical Weapons, has been established to inspect suspicious activities in signatory countries.[37]

The Convention directly affects a large number of private companies which have no history of involvement in military programs. First, many chemicals routinely used in commerce have been identified in the Convention as also having military potential and are therefore subject to international accounting requirements carried out by the companies and to international inspections of the associated manufacturing facilities. Second, countries that do not join the Convention will be penalized economically in that their companies will be unable to import from signatory countries those chemicals on the control lists, which are quite lengthy. Denial of access to chemicals produced in the member countries would surely crimp plans of nations interested in developing and maintaining a modern chemical industry.[38]

The most important American chemical industry group, the Chemical Manufacturers Association, strongly supported the Convention using the following logic articulated by its spokesperson:

> The *Chemical Weapons Convention* is a good deal for American industry. It protects vital commercial interests. I know because we helped design the reporting forms. And I know, because we helped develop inspection procedures that protect trade secrets while providing full assurance that chemical weapons are not being produced. The Convention makes good business sense and good public policy.[39]

Despite this strong industrial support and a vigorous government campaign to gain ratification of the treaty in the U.S. Senate, ratification was a difficult process and reflected concerns not only of the United States but of other countries as well. The arguments against the Convention included the following:

- Russia will not be able to destroy its stockpiles due to the high price tag.
- Some countries of greatest concern—Iraq, for example—will not join because of their interest in retaining the option of chemical weapons.

- Inspections of those that join are not foolproof, and cheating will certainly be easy.
- Another new international bureaucracy is unnecessary.
- If the United States signs this imperfect agreement, we will let down our guard against terrorism, thinking that the problem has been solved.[40]

However, in a most eloquent defense of the convention which undoubtedly swayed some reluctant senators, Secretary of State Madeleine Albright underscored that the Convention would set a standard for judging international behavior. When nations do not meet this standard, they will be ostracized from the international community and treated accordingly; in the absence of such a standard, we would have much less leverage in discouraging dangerous adventurism and in coping with international terrorism, she argued.[41]

After heated Senate debates, the treaty was ratified in the spring of 1997. While some countries may not join, more than 100 will, proclaimed the jubilant advocates. Then they added that imperfect inspections are better than no inspections. The United States became a charter member as the convention entered into force in May 1997.[42]

Another stream of international cooperation to help stem the tide of terrorism has been the deliberations of the Australia Group. This *informal* gathering of representatives from thirty of the principal industrial countries, established in the early 1980s, meets periodically to review national policies concerning exports of equipment and technical data used in the production of chemicals that have not only commercial uses but potential application in producing chemical weapons as well. Much effort is devoted to persuading all members to adopt comparable approaches for limiting transfers of sensitive items to irresponsible customers.[43]

In short, the major pieces of an international watchdog regime are in place for encouraging nations around the world to block sales of sensitive chemical items that could end up in the hands of countries or groups with hostile motivations. This regime complicates the task of terrorists who would like to purchase chemical agents or manufacture agents themselves. Of course these arrangements will have more sway in deterring the production of chemical weapons on a large scale than preventing illicit arrangements in small facilities that do not depend on elaborate equipment.

In the United States, as in all signatory countries, new regulations are required to carry out commitments of the treaty. The United States has not yet enacted the necessary laws for promulgating such regulations. Fortunately, laws and regulations governing the handling of many potentially dangerous pharmaceuticals, pesticides, and industrial chemicals are already on the books. They have sensitized companies to the importance of knowing in detail the characteristics of their chemical activities, and the new law should be enacted without too much difficulty.

In this regard, in 1973, the EPA recruited me to lead the agency's effort to develop its initial program to regulate industrial chemicals. The first step was to compile a comprehensive inventory of all chemicals being produced and used in the United States. For four years, I was the point person for the governmental effort to put into place the *Toxic Substances Control Act*. This law, enacted in late 1976, provides EPA with the regulatory authority to require such an inventory. Also, EPA may investigate and, if necessary, control any one of the industrial chemicals that are being bought and sold in the United States at any time. The objective is to reduce the likelihood that Americans will be exposed to dangerous levels of toxic chemicals while at the same time not unnecessarily burdening American industry with reporting, testing, and other regulatory requirements. A large number of these same sixty-five thousand chemicals now find themselves on the new lists of chemicals of concern to the international inspectors of compliance with the *Chemical Weapons Convention*.

When the legislation was enacted in 1976, the national interest in controlling toxic substances had little to do with terrorist misuse of dangerous chemicals. Rather, the focus was on the inadequacy of American industry's control of environmental releases during production and distribution of potentially dangerous chemicals, particularly those used widely throughout the economy. At the time, some environmental extremists referred to chemical industry executives as industrial terrorists. I spent much of my energy curtailing destructive practices that led to such charges—for example, industrial plant contamination of cattle grazing areas in Louisiana, lead poisoning in Dallas and arsenic poisoning in Tacoma, both from poorly operating smelters, and fish kills from industrial discharges of mercury in Tennessee and from sewage discharges of organic chemicals in New York.

However, these chemical calamities were not motivated by political agendas or hatred; they were the result of shortcuts to save money. Even the *midnight dumpers* who left toxic chemicals in shopping mall parking lots were not trying to make political statements, only trying to dispose of the waste at no cost.[44]

Of course, the most immediate problems were burns from leaking acids and asphyxiation by noxious vapors. New legislation reinforced regulatory programs that were already in place to help ensure that companies were effectively avoiding such mishaps.

For two decades, industry, like it or not, has been developing and sharing with the government much of the information about chemicals that should be kept out of the hands of terrorists. The available data repeatedly highlight the dual-use character of many toxic chemicals—chemicals crucial to responsible industrialists but hazardous to the public when in the hands of irresponsible parties. Also, chemical manufacturers have learned the importance of knowing their customers lest they end up in law suits based on the misuse of chemicals that originated in their facilities. While the *Toxic Substance Control Act* and related laws were never conceived as antiterrorism tools, we would certainly be far more vulnerable to chemoterrorists were such legislation not in place and working.

During my subsequent watch as the Director of EPA's research laboratory in Las Vegas, Nevada, I learned about another aspect of controlling chemical agents. There were frequent accidents on highways, railways, and waterways in which dangerous chemicals spilled into the environment. Emergency response was one of the missions of the Las Vegas laboratory. Accidents will continue to be common occurrences. With thousands of tons of chemicals in transit at any given moment, spills from transportation accidents are inevitable. Also, some chemicals are so unstable that a severe jolt as they move through populated areas can cause an explosion, threatening local residents and passersbys. Once in a while, major accidents in harbors, along rail lines, at pipeline breaks, or on roadways are disastrous and may require evacuation of industrial or residential areas or may contaminate food supplies, waterways, or drinking water supplies.[45]

The fire and police departments of most American cities, in cooperation with the chemical industry, are now well prepared to handle such accidents. Although confusion sometimes reigns as to the character of the chemicals involved, those in charge usually make sensible decisions in addressing such immediate issues as: Should water be sprayed on the spill? How dangerous are the fumes? What are the decontamination and cleanup procedures?

I gained great respect for the preparedness of the Barstow, California, police when they called me in the middle of the night to report that a truck from our EPA laboratory overturned and littered the highway with unmarked containers, oozing chemicals. They had responded with vigor. As two minivans with chemical response experts raced the two-hour drive from Las Vegas to the scene of the accident, the local police maintained perfect order in keeping traffic moving and cordoning off the area while preventing any tampering with the spilled cargo.

These same local responders from police and fire departments will be called upon to deal with terrorist releases of chemicals in our cities. Their base of experience in competently handling minor chemical incidents, and occasionally major mishaps, is already substantial. They appreciate the importance of immediate response to chemical releases; they respect the hazards in coping with unidentified chemicals; and they know how to find expert help when needed.

However, the step from responding to accidents to responding to planned chemical assaults is a giant one. Terrorists will probably make life as difficult and dangerous for the responders as possible, and contaminated areas may be measured not in terms of city blocks but in terms of square miles. Further, responders will be dealing with a crime scene that is still active—contaminated, perhaps subjected to delayed chemical releases, and possibly even booby-trapped. No local agency anywhere in the country is adequately prepared to respond to concerted efforts of major terrorist organizations, as illustrated by the simulated sarin attack on the most prepared city in the United States: Washington, D.C.

Recognizing the need for extra precautions in coping with terrorists who have access to dangerous chemicals, the FBI and several other agencies established a special chemical analytical facility in Atlanta at the time of the 1996 Olympic Games. The laboratory was immediately put to the test in analyzing soil and other samples following the explosion in Centennial Park. But even the best laid plans and intentions cannot always yield results quickly enough. As was the case in Atlanta, it may take hours to collect, transport, and analyze substances; but if the laboratory had uncovered signs of a suspicious chemical agent, the delayed results would have been useful in pursuing the perpetrator or in treating victims.[46]

The FBI has taken the lead in setting up similar laboratories behind the scenes of inauguration ceremonies and even the Super Bowl. It has organized much of these chemical response activities around the capabilities of its central scientific laboratory. The plan is to have eventually a new research and training facility in Quantico, Virginia, that can serve all agencies.[47] Meanwhile, there are so many high visibility targets that could attract terrorists in the United States that portable laboratories simply cannot be routinely deployed at remote sites.

The U.S. Army has issued a handbook *Medical Management of Chemical Casualties* for use on the battlefield and in case of emergencies at civilian arenas.[48] The handbook sets forth the signs and symptoms of major groups of chemical agents, including physical effects; techniques for detecting the presence of the chemical; decontamination procedures; treatments, including antidotes if appropriate; and protective gear needed during detection and decontamination. This excellent primer summarizes the characteristics of nerve agents. It is available in cyberspace to help counter the Internet classrooms for the chemoterrorists.

Responders are gradually being equipped to deal with chemical assaults, but as I have already suggested, terrorists are designing ways to sabotage the efforts of fire and police departments. In the words of the executive director of the International Association of Fire Chiefs:

> I am sure that our fire and emergency service leaders are not aware of the fact that in nearly 50 percent of terrorist bombings, a secondary bomb (often larger than the initial bomb) has been

set to attack the responding rescuers. We can be sure that terrorists are studying the TV videotapes of incidents to understand
how we respond with law enforcement, fire, and emergency
medical services. They are looking for weak areas that they can
exploit in future incidents. And, these are not old-fashioned anarchists; today's international terrorists have proven themselves
to be very intelligent, well financed, up on the latest electronic
technology and willing to sacrifice themselves for their cause.[49]

A chemoterrorist incident followed by a bombing of the rescuers is indeed a frightening scenario to fathom. A bombing followed by a chemical attack on the responders would be even
more chilling.

Chemoterrorism undertaken either at the instigation of a
hostile country or by an independent terrorist group should be at
the top of the list of national security threats. The international
community has outlawed chemical weapons. Any employee of
any government or any other person involved in the planning or
carrying out of a chemical attack on anyone—armies or civilians—or even threatening such an act is not only a terrorist but a
superterrorist.

U.S. experts emphasize that chemoterrorism from abroad is a
diplomatic problem, an intelligence problem, a law enforcement
problem, and a military problem.[50] In this context, effective functioning of the *Chemical Weapons Convention*, collaboration among intelligence services, and restraint by all countries in exports of
sensitive items in accordance with recommendations of the Australia Group—backed by resolve to use military force if necessary—
are very important in containing this unsavory terrorist threat.

Additional steps are promptly needed. Aggressive international inspections under the convention can help discourage establishment of clandestine laboratories and pinpoint them if they
exist. International pressure should be applied to countries who
have not yet signed the convention. Countries who do not join
should be treated as pariah states with the international promise
of military intervention to disrupt any preparations for chemical
attacks. As we have seen, of all the new forms of terrorism, releases of poison gases in the very near term are surely among the
most worrisome and probably the most immediate.

At the same time, given the availability in the United States of dangerous chemical agents or of at least the ingredients to make any desired mix, a terrorist group focusing on targets within the United States need not bother to try to bring dangerous chemicals or suspicious equipment across our borders. International supply lines simply are not necessary. Perhaps it will be basement laboratories where dangerous chemicals can be prepared that will represent the most difficult vulnerability to offset. Government officials are worried that large hardware stores provide a sufficient source of needed supplies, at least for crude devices. Senator John Glenn contends that a facility the size of a large room could turn out tons of nerve gas. Others share his opinion. They point to the small laboratories used by the Aum Shinrikyo to prove their point.[51]

The federal government and our states have laws controlling the manufacture, handling, and use of dangerous chemicals. Policing those facilities where chemicals are manufactured and used has been important for worker protection and the prevention of environmental contamination. Now, with the possibility of theft or deliberate misuse of chemicals added to the list of concerns, enforcement activities must become even more intrusive into the laboratories of America, much to the chagrin of legitimate researchers and industrialists. Small-scale poisonings and even limited releases of poison gases may not be stopped, but preventing massive uses of chemical agents must be a priority.

At the same time, we must be better prepared for the worst. A New York City official responsible for responding to both biological and chemical terrorism attacks emphasizes:

> There should be state-of-the-art chemical detection equipment in any major city. Massive decontamination facilities are needed. There is no facility now. Protective equipment and suits will be needed, but who knows how much equipment, suits, supplies, and people would be needed. This doesn't seem even to have been studied. All this unpreparedness is just asking for trouble.[52]

Aggressively enforced laws to prevent chemoterrorism and heightened preparedness to respond to incidents are essential. It will be a challenge to effectively implement such commitments in an age when chemicals are critical to every aspect of our standard of living and are readily available. Yet this is precisely the task before us. To forestall serious consequences, all elements of the sprawling chemical complex of the nation must cooperate with law enforcement agencies in a concerted effort to prevent the development of a new generation of chemical weapons right in our own backyard.

CHAPTER 4

Bioterrorism:
The Last Frontier

*We should be prepared for bioterrorism casualties in
increments of five thousand, recognizing that some
hospitals may close their doors to contaminated victims.*
Fire Chief, Arlington, Virginia, 1997

*If people don't report their initial symptoms from a
disguised terrorist attack to a central organization, the
disease is just going to sit there and cook until the
outbreak is out of control.*
Chief of Epidemiology, Texas Department of
Health, 1997

*Any country with a modern scientific program can
produce efficient biological agents.*
Iraqi Army General Staff Training Manual, 1987

On Thursday, April 24, 1997, just one mile from the White
House, Scott Circle played reluctant host to a fleet of ambulances, fire trucks, police cruisers, and FBI vans that would remain there for hours. The sidewalks bustled with law enforcement personnel, flanked by reporters and photographers jostling for position. Police had cordoned off a two-block areas and firefighters wearing hazardous protection suits were deployed. One of Washington's main thoroughfares, Massachusetts Avenue, was closed at the busiest time of day. This was no fantasy of a tabloid reporter nor a practice drill of responders becoming accustomed to recently

acquired HazMat suits. It was a serious response to a real threat of bioterrorism.

Just off the circle, health officials had quarantined 100 people inside the B'nai B'rith international headquarters. A package leaking a foul-smelling red liquid had been discovered in the mail room. A broken petri dish in the 8-by-10-inch package was marked "antracis yersinia." Bacillus anthracis is the bacterium that causes anthrax, and Yersina pestis causes bubonic plague—both deadly diseases. The authorities were taking no chances that the newly arrived microbes would spread beyond the confines of the building.

Meanwhile, the air conditioning systems in nearby office buildings were turned off, lest they inadvertently suck in and circulate infectious microbes. Just off the circle, eighteen police officers, firefighters, and other on-scene workers had stripped to their underwear and were being hosed down with decontaminants.

Finally, by eight o'clock PM, the package had been examined, at least in a preliminary fashion, at the Bethesda Naval Hospital. The laboratory staff was satisfied it contained nothing infectious. Health officials informed the two B'nai B'rith employees under observation at the hospital and the other employees at the headquarters building that they were out of danger and could go home.

Through it all, anxious officials in the White House monitored developments but apparently were content to let the local responders handle the situation.[1]

Three days later the FBI reported that agar, a red gelatin-like material, was in the package. They explained agar was a harmless and readily available material derived from algae and used as a medium for growing bacteria. They added that the agar contained bacteria commonly found in homes, offices, and shopping malls. The FBI explained their public statement was delayed by tests that required incubation periods.

According to the FBI report, a group called the Counter Holocaust Lobbyists of Hillel took credit for the incident. The accompanying note called the apparent hate package a chemical weapon. The note derided Jewish liberalism and denounced the Jewish community in general.[2]

In the aftermath of the incident, the police department acknowledged the inadequacy of training for this type of incident, and the fire department expressed frustration over a lack of proper equipment. The fire chief apologized for any public embarrassment caused by hosing nearly naked personnel in full public view and promised to provide a tent the next time.[3]

There may soon be a next time. While psychological chaos from the specter of anthrax was the intent at Scott Circle, the use by terrorists of real anthrax spores—whether obtained directly from infected animals or purchased from suppliers of microbes to scientific laboratories—has become a possibility. Thus, understanding the characteristics of this bioagent is critically important.

An industrial city on the other side of the world provides a controversial yet informative source for the chilling details.

> Americans are simply wrong when they say that the anthrax poisoning that killed seventy inhabitants of Sverdlovsk in 1979 was the result of a release of bacteria into the atmosphere from a military microbiology research facility in our city. The information presented by American specialists is not accurate. It is based on their misguided political efforts to prove that we violated international agreements rather than on an objective scientific analysis of the situation. The deaths resulted from consumption of contaminated sheep and cattle. We had previous problems with natural contamination of our livestock, and so we are familiar with this possibility.[4]

In 1992 and again in 1994, a leading ecologist of Sverdlovsk (now renamed Yekaterinburg) in the Urals region of Russia used this argument to try to convince me the new Russia was not trying to cover up mistakes of the Soviet Union as alleged by other visiting Americans. While he shared my much greater interests in nuclear and chemical pollution problems of the region, the focus of my visits there, the ecologist's heated denials when discussing the anthrax incident with other Americans had made him so defensive that the topic inevitably colored our meetings as well.

This academician and some other Russian experts rejected the conclusions of a careful, American-led investigation organized by a Harvard professor in the 1990s to examine the release of the potent contaminant in his city. While I did not participate in this investigation, I had followed for more than a decade the disagreement over the origin of the anthrax contamination—a dispute that continues to spark international controversy even today.[5]

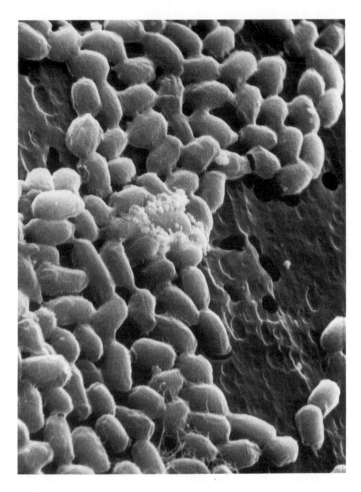

7. Bacillus anthracis (anthrax spores), above, and Yersinia pestis (plague bacteria), opposite page, may be the bioagents of choice of terrorists.

In 1992, Boris Yeltsin acknowledged the Soviet Union had conducted research prohibited by the 1972 *Convention on Biological Weapons,* and he vowed that they would cease such experiments. The U.S. government is convinced he was referring not only to biological experiments that led to the Sverdlovsk incident but also to other illegal activities.[5] The U.S. view, shared by almost all experts outside Russia, contends that an aerosol cloud of anthrax spores was accidentally released from a military high-hazard laboratory facility conducting convention-prohibited research.[6]

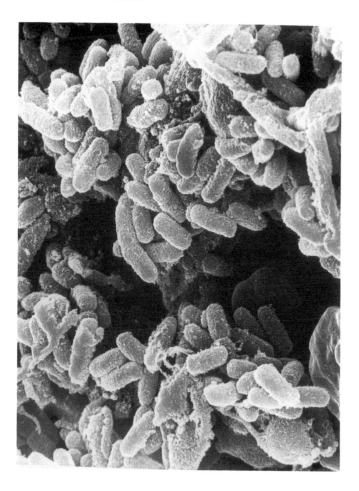

The results of the American-led scientific team supported this contention. The team conducted interviews with friends and relatives of victims in Sverdlovsk. They pinpointed the victims' locations at the estimated time of the spores release. They then noted the victims' positions relative to the direction of the wind away from the military base. The correlations appear to show conclusively that the anthrax outbreak could be traced to the microbiology institute at the base. At the same time, the results of a Russian investigation into the incident ordered by President Yeltsin in 1992 have yet to be disclosed to the public.[7]

Regardless of whether the contamination was the result of military adventurism or the vagaries of nature, the health toll was substantial. The Russian interviewees reported to the American researchers that they had suffered from the following maladies, in descending order of frequency: fever, shortage of breath, coughing, headaches, vomiting, chills, physical weaknesses, abdominal pains, and chest pains. The average hospital stay was one to two very painful days for the seventy who died and three weeks of misery for survivors.

Once the alarm was sounded, the actions of local authorities indicated much confusion among responders as to the source and method of dissemination of the deadly bacteria. On the one hand, local fire brigades washed down trees and workers quickly put down asphalt on several unpaved streets, suggesting remediation actions to prevent recirculation as airborne contamination. At the same time, officials confiscated and burned meat entering the city. They also took biopsies from sheep and cattle for laboratory testing, indicating a genuine concern over agricultural sources of anthrax. Meanwhile, kitchens and sick rooms throughout the south side of the city were disinfected, and a voluntary immunization program was launched with 80 percent of the fifty-nine thousand eligible people being vaccinated at least once. While multiple vaccinations over many months are required for complete immunization, such one-time emergency inoculations were intended at least to reduce the risks should anthrax spores still circulate in the air or be present in the food supply.[9]

This account of a two decades-old occurrence in a distant corner of the world might at first glance seem irrelevant to Americans. However, as I have noted, anthrax is a likely bioagent of choice by future terrorists. Unfortunately, the bacteria are relatively easy to obtain from infected cattle or sheep or from other sources. They can be made to multiply rapidly in small biological laboratories. The spores remain stable for a period sufficiently long enough to allow their dissemination onto distant targets. And, as seen, the effects of anthrax can be excruciating. While the disease is not contagious between humans, dispersion of bacteria through the air or through ingestion can pose fatal consequences.

Further, as indicated in Sverdlovsk, uncertainties will arise as to whether the source is nature or human, should terrorists seek to disguise their intentional dissemination of the dangerous pathogen. Should anonymity not be a priority, then blatant spraying of anthrax spores from aircraft has often been mentioned as a delivery method. As demonstrated, shipping bioagents in packages is obviously on the minds of some terrorists who would be satisfied with infecting a limited number of innocent recipients in order to make their point.

That said, the effective delivery of anthrax spores *in large quantities* onto a significant population, through spraying from aircraft or other means, is not an easy task. Simple crop dusting systems won't work. Fertilizers and pesticides are sprayed from planes in relatively large globules that promptly "sink" through the atmosphere onto specific fields. Bioterrorism agents should be released in smaller, dust-like particles that "float" in the atmosphere until someone breathes them. They must be released at the proper height above the ground; if they are scattered at too high an altitude, they will disperse before they reach nose level. Unfortunately, the Iraqis apparently have made progress in achieving such floaters.[10]

The extensive effort of the Iraqi government to build up their capacity to develop, package, and use biological weapons is the most obvious case in point in comprehending the opportunities and limitations in pursuing this genre of terrorist threat. During the 1980s, development of biological weapons was on a parallel but much less expensive track with development of nuclear weapons. UN reports indicate that Iraq was prepared to use whatever biological weapons it had ready should there have been a nuclear attack on Baghdad.

Had the Gulf War continued and had Iraq resorted to biological agents, there might indeed have been a nuclear attack on Baghdad, particularly by Israel if Tel Aviv became coated with deadly viruses or bacteria. Fortunately, biological agents were not released in the Gulf War, but Iraq's extensive development effort has given new reality to the concept of biological weaponry.

In March 1996, Ambassador Rolf Ekeus, the senior UN official in Iraq, reported:

> We should recall that in January 1991, just days before the land war broke out in the Gulf, Iraq had a biological weapons program which had produced a large amount of biological warfare agents. They had filled them in missile warheads. The warheads had been moved out to the active missile force. So they were fully deployed for use. There were twenty-five such warheads

filled with anthrax, botulinum, and aflatoxin. They had a number of biological bombs, that is 150 to 200 bombs which were brought out to two of the air bases. They were deployed and on complete alert. Iraq still may keep some biological warheads— missile warheads with biological agents or empty warheads together with biological warfare agents capable of being filled into these warheads. *We are concerned that at least six missiles, maybe up to fifteen or sixteen, are not accounted for.* And if they now have a missile that has a range of six hundred kilometers and biological warheads, we have a problem.[11]

My subsequent discussions with UN inspectors suggest the ambassador's report missed a few important details as to whether the Iraqi biological bombs would have actually worked. First, there is the previously mentioned problem of having systems which disperse anthrax spores or other agents in very small particle sizes that will float long enough until they make their way to the inhalation tracts of victims. While both Russian and American inspectors believed that the Iraqis could probably dispense particles sufficiently small, such dispersion systems were never adequately tested to prove the point. A related concern was the thickness of the shell casings within which the bacteria was packed. The inspectors considered that the force needed to open such a thick casing would disrupt much of the dispersal power of the rotary device within the casing.[12] However, it would be foolhardy to assume that Iraq's adversaries would be saved by malfunctioning technologies. And now six years later, the probabilities of such malfunctions have undoubtedly decreased.

In the 1970s, the Iraqis began procuring the equipment, technology, and materials that provided the infrastructure for their weapons program. According to U.S. intelligence sources, the Iraqis then spent about $100 million from 1980 to 1990. They acquired vaccine production plants from France, fermenters and laboratory facilities from Germany, agricultural spraying equipment from Italy, and bacterial culture samples from the United States, France, and Great Britain. UN reports indicate keen Iraqi interest in many infectious diseases including gangrene, tetanus, cholera, tuberculosis, and plague. It is easy to ask why didn't the U.S. government stop this activity? But at the time, disrupting the Iraqi buildup simply was not a priority.[13]

During the 1980s, the Centers for Disease Control, now named the Centers for Disease Control and Prevention, in Atlanta sent to Iraq, allegedly for use in studying public health problems, dozens of biological strains. Could any of these strains have been diverted for use as starting material for weapons? Other strains were sent to that country by the American Type Culture Collection, a private nonprofit scientific organization in Rockville, Maryland, that takes orders and sells cultures of microorganisms throughout the world. Such deliveries were made in full compliance with American export control laws that have subsequently been considerably tightened to prevent uncontrolled shipments of this type to Iraq or other rogue countries. The exports from both of these organizations had ceased by the beginning of the 1990s as the unacceptable behavior of Iraq became clear to all. At the time of the invasion of Kuwait, Iraq was producing and stockpiling agents at a dozen sites.[14]

As to further details of Iraq's biological program, UN inspectors have often remarked that the presence of dead dogs indicated suspicious activity. Dogs apparently were their favorite animals for laboratory tests of virulent agents; and donkeys, monkeys, sheep, rabbits, and guinea pigs were also in their testing corrals. Rumors circulated that a diarrhea agent was under development, although the source of the frequent internal discomfort of visiting foreigners was never shown to have a definitive connection to a germ warfare laboratory.[15] Finally, the Iraqis were not concerned about containing in tightly sealed laboratories the dangerous microbes that were under study, as is the standard practice. They simply vented the contaminated air out into the desert where the microbes would fade away or at most attack an unlucky nomad who happened to wander by at the precise moment of the release.[16]

Following the temporary expulsion of American-nationality UN inspectors from Iraq in late 1997, the Pentagon rightfully filled the airways with doomsday scenarios if the Iraqis were allowed to continue their biological programs. The media quickly picked up this theme. One report, based on interviews with a dozen inspectors, provided additional insights into the Iraqi program:

> The program included production of 528 gallons of the carcinogen aflatoxin mixed with chemicals for riot control. UN officials speculated that Baghdad's aim was to spray the chemicals on the

Kurds, producing an untraceable spike of cancer years later. Iraqi biologists also studied how to use bacteria to destroy vital crops and animals, including a pox that kills camels. Although U.S. officials were aware that Iraq had weaponized anthrax and botulinum toxin, they had underestimated the amount of anthrax by a factor of eight and the amount of botulinum by a thousandfold.[17]

The size of Iraq's scientific work force dedicated to biological weaponry remains unknown. The Iraqis have claimed at times that as few as ten researchers participated in the program, while other estimates indicate more than 100 scientists.[18] In any event, the pool of talent devoted to the effort was substantial. Many microbiologists took their training in British veterinary schools. In addition, Iraq developed its own advanced training programs.[19]

Finally, insofar as foreign governments were involved in the program, European firms claim Iraqis misled them to believe that dual-use equipment was for peaceful purposes. However, since Western personnel were on site to oversee the construction of plants and to assist with start-up procedures, they should have been in a good position to assess possible intentions. German companies, in particular, have been the focus of criticism for their participation. The German government has stressed that there was no real proof usable in court to show complicity of individual companies.[20] (See Table 6 for an estimate of the Iraqi arsenal.)

UN inspection procedures to monitor developments in Iraq have always been of questionable reliability. Indeed, it is difficult to envisage any procedures that would absolutely ensure no unauthorized activity is taking place in small hidden laboratories or in converted agricultural supply factories. Now, with pauses and delays in the inspections, the Iraqis have regained time to revive dormant programs in secluded enclaves, should that be their plan. Inspection procedures cannot confirm reconstituted facilities do not exist. Still imperfect inspections are better than no inspections as a supplement to other intelligence-gathering techniques.[21]

Even if the Iraqis were honest in their contentions that they have destroyed all biological weapons,[22] the country's capabilities to restart, at any given moment, their dormant programs within a matter of months or to assist other countries or terrorist groups to launch bioattacks are substantial. The most important resource, Iraqi scientists and engineers, are still in Iraq.[23]

TABLE 6.   Iraqi Biological Warfare Program

Agent	Declared Total Amounts
Anthrax	85,000 liters
Botulinum Toxin	380,000 liters
Gas Gangrene	3,400 liters
Aflatoxin	2,200 liters
Ricin	10 liters

Note: UNSCOM estimates are larger but not confirmed.

Deployed Delivery System	Anthrax	Botulinum Toxin	Aflatoxin
Missile Warheads	5	16	4
R-400 Aerial Bombs	50	100	7
Aircraft Aerosol Spraytanks	4		

Note: Iraqis claims these have been destroyed but UNSCOM cannot confirm.
Source: *Iraqi Weapons Mass Destruction*, February 13, 1998. Released by CIA in February 1998.

Beginning many centuries before the recent crisis over Iraq's biological weapons, such agents have been used in warfare as highlighted in Table 7. Despite the heartless nature of these acts, it would seem that the repugnance of deliberately spreading disease has not fazed some military leaders.

Twenty-five years ago, the World Health Organization (WHO) was so concerned about the possible debut of a new generation of lethal biological agents released on a massive scale that it assembled a team of international experts to quantify the problem. The group described how seven potent biological agents could wreak havoc should they be disseminated effectively (i.e., as floaters blanketing the target populations) from airplanes on cities of five hundred thousand people. The experts estimated the death tolls and the number of incapacitated would range into the hundreds of thousands. (See Table 8.)

Thankfully, a major bioterrorist attack has not been carried out. At a special school in Berlin during the 1980s, the former East German government trained Iraqi terrorists in the use of biological agents however. Possibly maps of targets within the United States graced the walls of the war rooms of these Middle East bioterrorists.[24]

TABLE 7.   Use of Biological Agents in Warfare

- During the fourteenth century siege of Kaffa, attacking Tatar forces catapulted plague-infected cadavers into the city.
- In 1763, the British gave blankets that had been used in a smallpox hospital to Native American Indians.
- During World War I, German covert agents spread infections to disrupt exports or damage local operations as follows: sheep in Romania, mules in Mesopotamia, horses in France, livestock in Argentina, and animal feed in the United States.
- During World War II, Japanese military forces contaminated Chinese water supplies and food items with several agents and disseminated fifteen million fleas, infected by plague-laden rats, over several Chinese cities.
- During World War II, the Germans polluted a large reservoir with sewage.
- In the 1960s, the Viet Cong smeared pungi sticks with excrement.

Source: George W. Christopher, Theodore J. Cieslak, Julie A. Pavlin, and Edward M. Eitzen, Jr., Biological Warfare, A Historical Perspective, *The Journal of the American Medical Association*, August 6, 1997, pp. 412–417.

TABLE 8.   Fatalities from Dissemination of Bio Agents Upwind of an Urban Center of 500,000 People

Disease	Number of Dead/Incapacitated (from 50 kilograms)
Rift Valley Fever	400/35,000
Tick-borne Encephalitis	9,500/35,000
Typhus	19,000/85,000
Brucellosis	500/100,000
Q Fever	150/125,000
Tularemia	30,000/125,000
Anthrax	95,000/125,000

Source: George W. Christopher, Theodore J. Cieslak, Julie A. Pavin, Edward M. Eitzen, Jr., Biological Warfare, A Historical Perspective, *The Journal of the American Medical Association*, August 6, 1997, p. 415.

There have also been small-scale attempts by American extremists to disrupt life in the United States through implanting dangerous pathogens, often in food and water. (See Table 9.) These efforts illustrate one type of what might be called primitive bioterrorism that is already *past* our doorsteps.

In the years ahead terrorist groups somewhere in the world will surely attempt to trigger disease epidemics—acts of superterrorism. The bioagents developed long ago are the most likely candidates for such schemes.

TABLE 9. Incidents in United States Involving Biological Agents

- 1972: Members of a U.S. fascist group called Order of the Rising Sun found in possession of 30–40 kilograms of typhoid bacteria cultures with which they planned to contaminate water supplies in Chicago, St. Louis, and other large midwestern cities.
- Members of an Oregon cult headed by Bhagwan Shree Rajneesh cultivated Salmonella (food poisoning) and used it to contaminate restaurant salad bars in an attempt to affect the outcome of a local election. Although 750 people became ill and forty-five were hospitalized, there were no fatalities.
- 1995: Two members of the Minnesota Patriots Council, a right-wing militia organization advocating violent overthrow of the U.S. government, were convicted of conspiracy charges for planning to use ricin, a lethal biological toxin. They have appealed the convictions. Douglas Baker and Leroy Wheeler, the two men, allegedly conspired to assassinate Internal Revenue Service Agents and a deputy U.S. marshall who had served papers on one of them for tax violations.
- 1995: Thomas Lewis Lavy of Arkansas was charged with attempting to smuggle 130 grams of ricin across the border from Alaska into Canada in 1993 and with intent to use the toxin as a weapon. The next day, Lavy hanged himself in his jail cell.

Source: Jonathon B. Tucker, Chemical/Biological Terrorism: Coping with a New Threat, *Politics and the Life Sciences*, September 1996, p. 170.

There is also concern that in future years, there will be an introduction of new agents, such as influenza viruses. In order to create vaccines and other antidotes, genetic engineers are mastering the intricacies of molecular structures that cause different types of influenza. Could they divert their work to devise means for deliberately causing massive deaths? It should not be forgotten that the 1918 flu epidemic killed more than twenty million people worldwide. We are only now learning how that virus is put together so we can prepare appropriate defenses should it reappear.[24]

Other scenarios predict that scientists could introduce manipulated genes into food supplies, genes that will render large quantities of foodstuffs inedible. Also as previously noted in our discussion on the World War II experience disenfranchised groups could contaminate livestock and agricultural products of companies they resent.[25]

We are frequently reminded that the Ebola virus could do us all in. While in years past Ebola was not viewed by military authorities as a virus likely to be used as a weapon, experts increasingly believe that it could be used by terrorists. Filmed expeditions to Africa have brought back scenes of the effects of the Ebola virus, which has caused eight hundred deaths during the past two decades. The press has reported that members of the Aum Shinrikyo were in Africa in search of the virus. While many biological agents incapacitate their victims, Ebola has a kill rate of up to 90 percent.[27] One description of the effects of Ebola is as follows:

> Within days of infection, Ebola patients suffer from soaring temperatures and excruciating joint and muscle pain. The throat is so sore that swallowing anything, including one's own saliva, is intolerable. The connective tissue liquifies. The skin becomes like soft bread—it can be spread apart with one's fingers and blood oozes out. Victims choke as the sloughed off surfaces of their tongues and throats slide into their windpipes. Every body orifice bleeds. Even the eyeballs fill with blood that leaks down the cheeks.[28]

In 1994, at the top of *The New York Times* best-seller list was *The Hot Zone*, billed as a *terrifying true story*. This book about the possible spread of infectious diseases plied the reader with gory scenes from hospital wards where patients succumbed to the rapid disintegration of their internal organs wrought by Ebola or closely related viruses.[29] Since viruses have been spread inadvertently from Africa, sometimes by infected international travellers, and from the Philippines by infected monkeys, the allure of these dangerous viruses for terrorists seems clear.

At the same time, we should not lightly dismiss the technical problems in handling an agent as dangerous as the Ebola virus. The Iraqi program indicates that bioterrorism, like chemoterrorism, is dangerous for both the instigators and the intended victims. Even if well-designed laboratories and protective clothing are available, the procedures to ensure personal safety are not learned overnight. Nor should we forget the difficulties encountered by the Japanese army during World War II. Though they were successful in developing a host of virulent pathogens, they could not solve the problem of effective dispersal on the Chinese or other target populations. They tried using explosives to disseminate the agents, but only a limited number of particles, small enough to breathe, were successfully released.[30] This failure provides at least some level of consolation.

In the plus column for terrorists, however, many technical difficulties of the past *are* being overcome. As demonstrated in Iraq, the pool of qualified personnel with relevant skills and credentials is growing; biotechnology equipment and supplies are increasingly available legally in international markets. Both factors broaden access to expertise and technology by rogue governments and terrorist groups. Also, though better microbiology research and production techniques are being perfected for useful medical and agricultural purposes, some could almost immediately be diverted for nefarious ends. Still, in the near term, small groups will have difficulty in incorporating biological agents into effective weapons on a significant scale without a government providing both technical resources and a cover of legitimacy for their activities. In rare instances, autonomous groups might possess the capability to acquire and deliver biological agents, as apparently was the Aum Shinrikyo's intent in their visit to Africa.[31]

Of more than a little concern are the many university, industrial, and governmental research laboratories in the United States that boast very sophisticated biomedical facilities. Although devoted to the public good, they can be relevant to emerging bioterrorism schemes. One possible scenario finds an international terrorist organization enlisting specialists at an American research center, directly or through a middleman, to support its activities. The American surrogates at the facility provide access to technologies and equipment for a fee, or perhaps even hire on as highly paid technical consultants. The overwhelming preponderance of American researchers are loyal professionals, but a few bad apples could have devastating consequences.

Another episode could feature a group of domestic dissidents adopting approaches that are more refined and on a larger scale than the primitive bioterrorism assaults of the past. They might attempt, for instance, to infect air conditioning systems, contaminate drinking water supplies, or toss plague-filled lightbulbs into subway tunnels.

We cannot rely on luck to stop malcontents in their tracks before such wild-eyed schemes materialize and wreak havoc. Therefore, wide-ranging strategies for detection and intervention must be a priority for America's law enforcement. However, given the seemingly unlimited possibilities for such disturbing scenarios, major stumbling blocks will hamper both the anticipation and countering of bioterrorism in the United States or elsewhere around the globe. There will be difficulties in predicting targets, problems in initially detecting attacks, and delays in responding before severe damage has taken its toll.

Airports, large buildings, sporting arenas, resorts, famous landmarks, and other locations attracting both crowds and the media are on the hit lists of terrorists. Should travelers or tourists unknowingly inhale or ingest odorless or tasteless agents, the effects of the silent enemies would probably be delayed for days. Initially, the effects could be confused with natural outbreaks of disease. If massive illnesses or deaths occur within a few days, the distinctions should become clear, since naturally occurring outbreaks usually build up over a period of weeks or months.

Fortunately, many difficulties in preparing for the worst have been recognized at the highest levels of the American government. Efforts are under way, in Washington and internationally, to reduce the possibility of bioterrorism and to improve capabilities to respond effectively in the event of an attack. American officials now recognize rapidly emerging biotechnologies promises to enhance the quality of life through advances in agriculture, medicine, and sanitation, but they may also have a serious boomerang effect.

My own professional experience with mosquitoes reveals some of the unusual connections between biological weaponry and legitimate activities. In fact, my introduction to the U.S. biological warfare program began innocently enough in 1970 during my visit to an African junkyard in Dar es Salam, Tanzania. At the time, I was an American foreign assistance official, accompanied by a WHO virologist, searching for mosquito breeding grounds.

I was armed with numerous inoculations to ward off numerous diseases of all types. At the junkyard, I had no difficulty finding mosquitoes. The virologist confirmed that among the many types of mosquitoes in this disgusting dump were *Aedes aegypti* mosquitoes, the carriers of yellow fever. But why was this particular mosquito, and not the other 200 mosquito strains found in Africa, of interest? Indeed, by 1970, the eradication of yellow fever in almost every country of the world had been one of the WHO's proudest achievements. Also, the notion of deliberately spreading diseases via mosquitoes had lost its popularity with the armies of the world, at least for the time being.

On this trip, I was not especially interested in yellow fever or in biological warfare. My mission was to organize a mosquito sterilization experiment on the *Aedes* as a stepping stone to using similar techniques for eradicating the *Anopheles* mosquito, the particular breed that carries the malaria vector. Malaria was and remains one of the most serious health problems in many African countries. The *Aedes* was easier to handle—to artificially breed and to capture—than the *Anopheles*. If sterilization experiments were not successful on the *Aedes* (yellow-fever carriers), there was little likelihood they would be successful on the *Anopheles* (malaria carriers). If they were successful, however, similar experiments would be undertaken with the *Anopheles*, with the hope of eventually controlling malaria.

The experiment would take place on a small island off the coast of East Africa, uninhabited by people but populated with multitudes of mosquitoes. Large numbers of *Aedes* that had been genetically altered by chemical techniques, would be released on the island. When they mated in the field with neighboring swarms, their offsprings would be born sterile, eventually taking the final buzz out of hordes of mosquitoes. Experts predicted the entire strain would soon be eliminated throughout the island.

The theory seemed plausible enough. The challenge lay in developing a way to rapidly breed huge batches of the genetically modified mosquitoes. Upon return to the United States, I posed the question to American scientists who promptly urged using the army's mothballed mosquito factory at Pine Bluff, Arkansas. This hasty interchange gave me my first clue that a breeding facility used by the U.S. Army's biological warfare command in the 1950s and early 1960s even existed. The Pine Bluff mosquitoes, it turned out, were originally destined to be infected with all manner of nasty diseases that would then be imbedded into their victims on the battlefield through seemingly innocent mosquito bites.

During the 1950s, American army researchers eagerly investigated techniques for infecting large colonies of mosquitoes with the yellow fever virus. They studied the feasibility of disseminating the insects from airplanes over enemy territory. Indeed, in 1956 the army released *uninfected* mosquitoes in a residential area of Savannah, Georgia. With the assistance of unsuspecting local residents, prepared to do anything to help get rid of the new pest infestation, the army conducted surveys to estimate the number of mosquitoes that entered houses and bit people. Larger-scale mosquito-dispersion tests were subsequently conducted on military bases in the search for the most effective ways to spread these vectors of yellow fever.[32]

Meanwhile, the army was also experimenting with mosquito-laden artillery shells for use on the battlefield. When the munitions reached their targets, they would split open, releasing vectors of stinging death. Enough of the mosquitoes could survive impact and deliver the selected disease via mosquito bites to enemy populations, so thought the military.[33]

Fortunately, by 1970 such plans had been cancelled, as the U.S. government led an international effort at that time to rid the world of biological weaponry. The specially designed cages, feeding trays, and infection devices were soon on their way to East Africa for my more constructive project. While hardly an antiterrorism project, it opened my eyes to the power of scientists to manipulate even the smallest species.

However, the experiments were never carried out. After supporting a token effort to breed and release a few mosquitoes in the field, the American foreign aid officials controlling the purse strings became impatient with a research program that could provide results only in the distant future. They diverted the $1 million originally earmarked for the mosquito experiments to programs for solving more immediate problems of food shortages. Twenty-five years later, eradication of malaria continues to elude the world's best scientists. But at least our mosquito factory has closed its doors to military purposes.

Our army, and presumably other armies as well, long ago ceased experiments to infect mosquitoes, fleas, rodents, or other fuzzy carriers of deadly diseases. Plans to enlist potato beetles or to prepare lethal cattle feed to destroy agricultural resources have also been abandoned by military strategists. Table 10 gives some indication as to the types of projects the army worked on and later abandoned.

TABLE 10.   Biological Agents Weaponized and Stockpiled by U.S. Army (Destroyed 1971–1973)

Lethal Agents[1]
Bacillus anthracis
Botulinum toxin
Francisella tularensis
Incapacitating Agents[1]
Berucalla suis
Coxiella burnetii
Staphylococcal enterotoxin B
Venezuelan equine encephalitis virus
Anticrop agents[2]
Rice blast
Rye stem rust
Wheat stem rust

Notes: (1) used as a weapon; (2) stockpiled but not used as a weapon.
Source: U.S. Army Medical Research Institute of Infectious Diseases, August 1997.

Modern permutations of these earlier schemes picked up by terrorists may now call for the stirring of virulent chemicals and biological agents in a single cocktail. With the compliments of the Internet, such a frightening concoction has been placed at the disposal of the new purveyors of international violence.

Meanwhile, to this day, the U.S. Army has continued research on biological warfare agents searching for *countermeasures* to combat the use of dangerous pathogens. The current military research program in the United States investigates the effectiveness of only *defensive* measures to be taken should military adversaries or terrorists use biological agents of any type to destroy populations. Almost all of the effort is carried out quite openly. Each year, hundreds of foreign visitors observe research experiments at Fort Detrick, Maryland, where such activity is concentrated. But there exists a fine line between defensive and offensive research. Some critics contend that by continuing to carry out defensive research and, demonstrating and publishing the results of such research, we are transferring our findings to lands where our expertise could at least indirectly benefit terrorists with different motivations for understanding these agents.[34]

In considering questions about the wisdom of sharing such research, it is quite a stretch to say that the spin-off benefits to terrorists from defensive research on biological agents at Ft. Detrick are so great that our army's research should be carried out covertly. Secrecy associated with these activities would surely raise many questions at home and abroad as to political intentions and hidden discoveries. Indeed, some foreign policy experts argue that, symbolically, the openness at Ft. Detrick is critically important in encouraging other countries to follow suit and *not* carry out their dangerous, but entirely legal, investigations behind closed doors.

Another line of criticism lodged at the U.S. government is that the Department of Defense is naive in building up the biomedical know-how of developing countries. For example, the Pentagon has established and supported for many years biomedical research facilities in tropical countries. In 1978, when I led a scientific mission to Malaysia for the State Department, it was immediately clear to me that the U.S. Army's biomedical program there was a crown jewel within the scientific establishment of the country. The program was carried out in close cooperation with Malaysian specialists at the Institute for Medical Research in Kuala Lumpur. Local researchers had made remarkable progress in investigating the causes, effects, and eradication of typhus, typhoid fever, leptospirosis, malaria, and other tropical diseases.

The Department of Defense appropriately justifies such programs as increasing the understanding of the characteristics of diseases American troops might encounter while serving abroad. In 1997, they still maintained six overseas laboratories. Civilian agencies within the U.S. Public Health Service also have supported biomedical research abroad in their efforts to understand the origin and behavior of infectious diseases which know no boundaries and can reach the United States.[35]

To suggest terminating biomedical research on tropical diseases caused by nature but which also could be triggered by terrorists seems outlandish, given the importance of eradicating these diseases whenever and wherever possible. These diseases claim the lives of tens of millions of people every year. Nevertheless, the skeptics point out the complexities of dealing with bioterrorism, a form of terrorism that twists the very technical tools designed to protect humankind into insidious weapons to destroy it.

Furthermore, even if it were desirable for the United States to contain scientific knowledge concerning the behavior of biological agents, our ability to do so would be limited. Too many scientists in too many countries work in biotechnology, and thousands more from all over the world are trained every year. The only realistic approach is to maintain full awareness of the potentially destructive use of biotechnology should it fall into the hands of terrorists and to contain such use whenever possible.

The international marketing of the equipment and strains needed to begin the process of cultivating viruses and bacteria seems to be more suitable for control. Materials and equipment in laboratory inventories, and even in transit, are easier to identify than knowledge buried in the heads of specialists. At the same time, nearly all of the materials and items of equipment used to cultivate biochemical warfare agents have legitimate uses in the production of beer, wine, food products, animal feed supplements, pesticides, vaccines, and pharmaceuticals. Fermenters, centrifuges, freeze dryers, and efficient air filters have multiple uses. Honest researchers have valid reasons for needing many items found on the shopping lists of would-be bioterrorists.

As to availability of strains, in the past when vendors received requests written on a university or research institute letterhead, whether it be for seed cultures of anthrax or for certain influenza strains, the principal issue often raised was the method of payment.

The responsible approach is not to cut off the flow of equipment and of strains crucial in the war against infectious diseases and agricultural pests.[36] Along with the efforts of governments to improve their balances of trade and of the private sector to turn profits, both government agencies and multinational firms in many countries must use greater discretion in screening their customers if the threats of bioterrorism are to be held in check. Indeed, responsible vendors are becoming much more careful, not only in complying with export control requirements, but also in knowing their customers at home and abroad.

During the past decade, the issue of terrorists using biological agents against American targets has gradually risen to the top of the national security agenda in Washington. There is widespread recognition that while the incidents to date have been limited in number and scope, the potential for major attacks is great and growing. In 1996, President Clinton signed two new laws of considerable importance. The *Anti-Terrorism and Effective Death Penalty Act of 1996* criminalizes the use of chemical or biological weapons and regulates distribution of pathogenic microorganisms and toxins. The *Defense Against Weapons of Mass Destruction Act of 1996* links domestic programs to combat terrorism with efforts abroad to stem the proliferation of weapons of mass destruction.

Whether the predicted source of a terrorist act be renegade students working in a university microbiology laboratory, extremist militia groups who have built their own high-tech facilities, or fanatics from abroad searching for vials of poison, preventive measures may demand considerable intrusiveness requiring the redefinition of civil liberties. Responses may demand an expansion of police powers to limit the damage.

Will the American public accept a heavy hand from the authorities? Probably not, if we are reacting to only a minor incident or two. If there is a major disease outbreak with thousands, or even hundreds, of deaths attributable to bioterrorism, widespread support for many types of new intrusions into everyday life may be much more likely.

We must also address the terrorist cookbooks on the Internet, using their relevance to bioterrorism as an excellent case in need of attention. While the new recipes may not have all the information that a high-tech superterrorist needs to go from a bacterial strain to an effectively disseminated aerosol, some can be of help to novices. More important, their unchallenged availability signals societal tolerance of terrorism as a legitimate option. The serious debate over pornography on the Internet pales in comparison.

What can be done? Isn't the Internet no more than an automated library at your doorstep? If so, can we really close down sections of libraries all over the world? One significant step has been a 1997 court decision regarding Paladin Press, publisher of the book *Hit Man* that was advertised on the Internet. The court ruled Paladin was not protected by the right of free speech and could be sued for damages based on use of the assault techniques described in the book.[37]

With this decision as a start, the time has come when free speech should no longer be the cover for disseminating dangerous material. The president should promptly establish a commission to assess the extent of the dangers posed by distribution of technical guides for terrorists via the Internet or other means. The commission should determine the range of legal and policy constraints that could be invoked to deter such unbridled dissemination of formulas for death, and recommend specific steps to address the issue. If you can be punished for calling fire in a crowded theater when no fire exists, you should be punished for disseminating the information on how to set a real fire in that hall.

At the same time, everyone welcomes the disaster-response handbooks, guidelines, and instructions rapidly being prepared by the federal government for use by local agencies. The U.S. Army's handbook, *Medical Management of Biological Casualties,* is a step in this regard. In understandable and precise language, it describes the recognizable effects of the most likely bioagents and the roles of vaccines, drugs, and other prophylactic and therapeutic approaches. Handy tear-out pages are devoted to anthrax, cholera, plague, tularemia, Q fever, smallpox, Venezuelan equine encephalitis, viral hemorrhagic fevers, botulinum toxins, staphylococcal enterotoxin B, ricin, and trichothecene mycotoxins. This is a spread we hope will never reach our tables, but we should be ready. Should there be bioterrorism attacks in the United States, the public's interest in the topic will increase dramatically. This information about precautionary measures, probably of little interest to terrorists, is now available via the Internet to all interested parties.[38]

Also, the federal government is expanding support of research to enhance our ability to respond. A number of federal agencies are intent on providing fire departments with new devices that can signal the presence of an agent, perhaps even identify the agent. This is not a simple task and the initial costs of even primitive devices that can detect a range of bioagents will cost tens of thousands of dollars. The goal is to increase, within a few years, the reliability of such devices as well as bring down their costs.[39]

The Department of Defense has taken the lead in expanding support for research to open new horizons for a long-term battle against bioterrorism. Its Defense Advanced Research Projects Agency, credited with being the originator of the Internet, has launched a program with initial funding of $50 million per year, which is scheduled to double by 1998. This is a sizable increase over most funding in this field thus far. Three of its early projects are designed to:

- Develop synthetic bioskins, filters, and other protective gear based on recently proven principles of cellular structure.
- Create novel, broad-spectrum antimicrobial agents.
- Compile a repository of intercellular antibodies that can be used in the development of new therapeutic agents.[40]

In support of public health aims and, by extension, countering terrorist threats, expanded worldwide efforts to develop a wider range of antiviral and antibiotic drugs for fighting outbreaks of dangerous pathogens should also be pursued. One noteworthy effort is a Harvard University's project to eliminate the lethal potential of toxins by combining them with other molecules that shore up the body's natural defense system. Scientists have already combined bits of anthrax strains with proteins that stimulate the body to fight everything from food poisoning to AIDS to cancer. If successful, the final products will be new vaccines of considerable significance.[41]

Meanwhile, the pharmaceutical industry conducts large-scale research on antimicrobial drugs. Some of these drugs should eventually be of help in responding to certain types of biological attacks, although remedies for a number of potential bioagents are not yet being pursued.

In time, vaccines that protect against multiple viruses and devices to sense the presence of even the lesser known viruses may become realities. For the near-term, we must use more limited technologies to the fullest when responding to potentially dangerous events.[42]

Whatever counterterrorism strategies the United States chooses to promote and support in the area of biological agents, international cooperation will be crucial. To its credit, the international community of nations has attempted for many decades to excise biological weapons from the inventory of accepted forms of conflict resolution.

In 1925 the United States and many other countries signed a protocol in Geneva that called for the prohibition of "the *use* of bacteriological methods of warfare."[43] In 1972, the United States joined other countries in signing the *Biological Weapons Convention* which forbids signatories "to develop, produce, stockpile or otherwise acquire or retain microbial or other biological agents, or toxins whatever their origin or method of production, of types and in quantities that have no justification for prophylactic, protective, or other peaceful purposes."[44] Finally, in 1975 the United States ratified these international agreements and formally renounced any intention to acquire a biological weapons capability. More than 100 other countries have also ratified these documents in an effort to provide an important legal framework for international efforts to prevent the introduction of biological weapons into the arena of conflict.[45]

In January 1998, President Clinton announced more vigorous efforts to seek worldwide agreement on a protocol that would provide for international inspections to help ensure compliance with these agreements. The United States had been hesitant due to concerns over compromising proprietary information of pharmaceutical companies, but procedures have been worked out so that American companies can rest easy that inspections could be effective without compromising their proprietary secrets.[46] Our Iraqi experience underlines the difficulty of catching cheating nations, regardless of the intrusiveness of inspection procedures. However, we have no alternative but to come at the problem from all directions. Informed experts have calculated our intelligence information on the activity in rogue nations relating to biological agents as two on a scale of one to ten, with ten being good, and international inspections should raise the reading on the compliance thermometer.[47]

Another international arrangement directed to preventing the possibility of biological weapons being produced and used came into being in 1990. Earlier, about thirty nations joined together in the Australia Group to limit exports of sensitive items that could assist rogue states or terrorist groups determined to use chemical weapons. In 1990, the Australia Group expanded into the biological arena. They have identified the many dual-use items highlighted earlier, and they are promoting consistent export control regulations to prevent such items from falling into the hands of irresponsible parties.[48]

These actions provide an important legal and administrative structure to complicate and even thwart terrorists intent on using the poor man's nuclear weapon. They have encouraged nations to adopt domestic legislation that regulates dangerous goods. However, given the dual-use nature of nearly every ingredient necessary for biological weaponry, additional international steps, such as trade sanctions, are essential to crimp activities of the twelve or so countries currently on watch lists, countries that may choose not to comply with any international agreements in this field.[49]

As to responding to incidents, a novel approach was adopted by the Swiss Disaster Relief Unit during the Gulf War. The unit established a team of biological warfare specialists (named Task Force Scorpio) and placed it at the disposal of the UN Secretary General to be activated in the event of an attack. The Task Force's mission was (and remains) to fly to the scene of an outbreak to identify the biological agent, prescribe the antibiotics and decontamination supplies needed, and determine measures to protect relief workers entering the area. Other nations should follow this excellent example of sharing the responsibility for fighting global terrorism.[50]

Finally, for public health reasons, long overdue is an expanded international effort to assess the significance of outbreaks of infectious diseases throughout the world. The WHO is the natural leader for such a program. The greater the effort in public health, the stronger will be the capabilities of the international community to respond not only to natural disease outbreaks but to those artificially induced by terrorists. At present, the greatest terrorist threat is probably from the use of agents already familiar to the scientific community. There is considerable understanding within the international scientific community of the health impacts of their use—details that are of great advantage when it comes to developing preventive measures and antidotes.[51]

One example of past international collaboration deserves special mention, namely the successful worldwide effort to eradicate smallpox. Considered a disease of the past, smallpox could, in the eyes of some experts, become a prime candidate as a terrorist agent of the future. Should all strains of smallpox now being held for research purposes be destroyed to lessen the threat of theft or diversion into the hands of terrorists? Many researchers argue for retaining the strains, pointing out the linkages between smallpox research and research on other poxes not yet eradicated. These voices also question whether all smallpox strains were really placed in one repository in the United States and one in Russia, as mandated by the WHO in the early 1990s. They worry that if we outlaw smallpox strains, only the outlaws will have the strains. At the same time, many governments want to get rid of the virus altogether. They are convinced that there are no clandestine repositories. They resent Russia and the United States serving as big brothers for the world. Meanwhile, some scientists contend that since the genetic sequencing is known, the virus can be reconstructed if necessary. The debate illustrates the difficulties in coping with the many facets of bioterrorism.[52]

While international cooperation will surely help in efforts to protect Americans at home and abroad, the importance of devising counterattacks to repel this new pernicious dimension of terrorism is clear. The challenges in discovering concealed terrorist laboratories and activities are daunting. International inspections and export controls that require identification of proposed end users are helpful but not perfect. Much will depend on tips to law enforcement—by defectors, infiltrators, and even passersby. Stronger steps can be taken to control strains earmarked for research. Laboratories, wherever they are located, should be certified as not only safe but reliable and operating for legitimate purposes before they are entitled to access to certain strains. Alarms should ring if an order for a dangerous strain comes from a questionable institution such as a research laboratory in Libya, an unknown midwestern laboratory—irrespective of the national or international stature of the scientists requesting the strain.

Further, the liability of those people involved in aiding or abetting the use of biological agents should be widely publicized. Whether they are government employees ranging from the head of state to the enlisted soldier, they are terrorists if they promote the use of biological agents or even threaten to use them.

Like the bartender suddenly responsible for coming between the last drink and the customer about to become a weapon of destruction on the highway, any person or any company providing dangerous biological strains and sensitive dual-use equipment should not be able to plead ignorance of past behavior of his or her customer. Suppliers must satisfy themselves, through documentation procedures, qualification measures, and recommendations from other suppliers, as to the legitimacy of their customers and the activities they pursue. Indeed, the scientists working in this field should develop their own international code of conduct and seek to enforce it through their professional and industrial societies. One renegade can damage the reputation of all, not to mention take the lives of many innocent people.

Finally, the vigor of the recent reactions of the U.S. government to the *possibility* of a biological attack sets an important precedent for other nations to follow. In recent years, the U.S. Congress defined as a federal crime virtually every step in the process of developing or acquiring a biological agent for use as a weapon. Congress has vested federal law enforcement agencies with broad civil and investigative powers to intervene before such weapons are used or even developed.[53] Our health departments, law enforcement authorities, and emergency response teams are rapidly improving their preparedness. Most important, terrorist groups with eyes on America should anticipate encountering swift and severe retribution for attempted attacks. Though a renegade group may declare its disinterest in public censure once the deed is done, the promise of wrath of unprecedented magnitude cannot be ignored—and the likelihood of severe consequences should have a deterrent effect.

That said, the most likely use of biological agents on a significant scale in the United States appears to be the release of a cloud of infected air, a biological aerosol, into a heating or air conditioning system or into a confined space bristling with people. Once released, it will be difficult to find the perpetrator.[54] Each year, the technical difficulties in coping with these or other types of bioterrorism increase. As the legal framework becomes stronger, it is up to all of us to alert authorities promptly should we encounter suspicious ingredients or sense strange activities in our neighborhoods. Our false alarm rate cannot be too high in dealing with viruses and bacteria that threaten human life.

CHAPTER 5

The Globalization
of Weapons Expertise

U.S. Pays Millions To Russian Scientists Not To Sell
A-Secrets.
The Washington Post, 1994

In 1997, 419 American visas were issued to Iranian
students attending U.S. universities and technical
institutions. A significant number are fanatical, anti-
American Shiite Muslims who provide a resource base
that gives Iran the capability to mount operations
against the United States.
Chief of FBI International Terrorism Section, 1998

Dissemination of some dual-use technologies that have
relevance to weapons of mass destruction needs to be
encouraged if populations in developing nations are to
improve their health, environment, and standard of living.
Former Congressional Office of Technology
Assessment, 1993

*I*t was October 30, 1996. It had been a long and difficult five years since the Soviet Union splintered into fifteen countries. As Vladimir Nechai, aged 60 and director of the nuclear weapons design institute in Snezhinsk, Russia, made his way home from work, he thought about how much had happened during that time. Normally invigorated by the chilly autumn air, Nechai took slow measured steps that night. His heart was heavy as he thought about Snezhinsk, a city in the southern reaches of the Ural

Mountains which for many years had been shrouded in secrecy. It had been identified to the outside world only by its post office designation of Chelyabinsk-70. Constructed with prison labor, the city was known for decades to only a limited number of Russians; no one else was to know of its existence.[1]

Of course, the leaders of the mighty Soviet Union had known about it and had always recognized the importance of the city and its devoted workers. As reward for keeping Mother Russia well stocked with nuclear weapons and related defense technologies, Moscow returned the favor by sending regular shipments of the highest quality goods available to the city. But, the smug pride the town's people once felt had turned to fear and desperation.

Better days were long past. Nechai had devoted most of his life to serving a communist government, which he passionately supported. That government was now discredited. He often retreated to drinking as a way to forget the degradation of his motherland in the eyes of the world. Meanwhile, economic hard times had engulfed this company town of seventy thousand people where one institute provided almost the entire income. Nechai had tried his best to maintain some semblance of work discipline at the institute. But the financial plight had taken its toll not only on Nechai but on five thousand scientists and ten thousand support workers. Their interest had always been in the mathematical equations and physics experiments that explain the phenomena of nuclear explosions, not on what had unexpectedly taken the place of their research: making enough money to put food on the table.

This day marked the four-month anniversary since most of the institute's staff had received their last monthly paychecks. Indeed, the standard of living in the city had been in sharp decline as Moscow systematically cut the institute's contracts. Despite five years of institute efforts to market their high-tech achievements for applications in civilian science, medicine, energy, and protection of the environment, few outsiders were ready to invest their financial resources in conversion projects that would pull the institute out of its financial tailspin. The institute was on the verge of a crash landing.

Nechai had two offices: one behind the security barriers of the institute's inner sanctum and one on the third floor of a building not far from his house in the residential portion of the town.

On this night, he ate a sparse dinner at home and retreated to his nearby office, moving into the study and closing the door. Minutes later, after writing a short note, he opened his desk drawer, pulled out a gun, and shot himself.[2]

Nechai's death caused quite a stir around the world. A few years earlier, a suicide death by a then respected Soviet nuclear weapons designer would have been unthinkable. Bomb building was one of the most honorable, and best paid, professions in the Soviet employment registry. Nuclear warhead designers were heroes—revered by the general population, highly esteemed by other scientists, and handsomely rewarded by the state. The creation and rigorous maintenance of the nuclear arsenal were patriotic tasks, reserved for the best of the Soviet technical elite. But times had changed.

Nechai's suicide had heavy implications. If a leading Russian physicist was so desperate over financial conditions to commit suicide, would others in similar circumstances be prepared to sell their secrets? Would Iraq, Iran, or other radical nations or groups searching for new weapons of terror offer at least temporary financial relief for the valuable knowledge carried in their heads? In 1991, Iraqi officials had reportedly approached Russian scientists about the possibility of buying a nuclear weapon for $2 billion. Although the offer was said to have been immediately rebuffed, as we have seen, the interest of Iraq and other rogue states in nuclear weapons clearly persists, and economic pressures and hardships in Russia have increased.[3]

I had visited Snezhinsk several years earlier. I had met Nechai, ever the charming host despite his problems. I was well acquainted with several of the director's close colleagues. However, until his death, the severity of the plight of the institute and its scientists was not obvious to me. Also, I suspect that Russian officials in Moscow had too readily assumed an unlimited resilience within Russia's scientific establishment to absorb an endless stream of both political and economic shocks. In true Russian tradition, local financial difficulties were well disguised from foreign visitors to the Urals. Despite frequent press reports about demonstrations by scientists in protest of delayed paychecks, the institute leaders' generous hospitality—exemplified by sumptuous banquets and plentiful drinks—convinced the casual visitor that the institute's reserves would see it through until relief arrived. The institute's willingness to allocate funds to the local branch of the outdated, but still active, local Communist Party indicated some change remained in the coffers.

If the truth be known, however, Nechai, in desperation, had obtained short-term loans from local banks to cover government shortfalls and to meet the payroll. With banks soon demanding immediate repayment of the ruble-equivalent of $5 million, he could no longer keep up the pretense.

A few hours after the scientist's death, a security guard who had been making his rounds became concerned, given the lateness of the hour, when the director had not re-emerged. The guard summoned the chief of security who found Nechai. As his final act, Nechai wrote that he wanted to be buried on the following Friday and that his unpaid back salary should be used to cover the costs of his wake. This former leader of the Soviet military complex, who would still be welcomed at physics institutions throughout the world, had become an insignificant figure as Russia staggered to regain economic viability.

Nechai's funeral added another somber chapter to the sad story engulfing tens of millions of his countrymen. The Russian government, possibly embarrassed by what they viewed as a betrayal of the state, was not represented at the funeral. Many of Nechai's friends were not present either, simply because they, the remaining elite of the nuclear weapons establishment, could not afford the airfare to travel to the Urals.

Local mourners were not surprised at the turn of events. They knew Nechai had been under tremendous stress trying to explain to the work force why paychecks had become an illusion. For years, he had been attempting to reassure the employees that times would get better only to repeatedly renege on his pledges and report to them on the continuing downslide of the institute.[4] "Suicide of Lab Director in 'Closed City' Underscores Angst," reported a correspondent of *The Washington Post* who visited the city following Nechai's suicide. To the potential detriment of world security, this angst among the nuclear weapons brain trust is not confined to one small city in the Urals.[5]

Given the desperate plight of many Russians, some of Nechai's colleagues may have become positioned for unanticipated career moves. As terrorism in distant lands reaches a new level, revolutionary and reactionary groups are definitely in a hiring mode, looking for the best technical skills they can buy. Fortunately, through 1997 there have been no public reports indicating that either rogue states or independent terrorist groups have been successful in enticing bomb designers to emigrate from Russia to work for them. Such irresponsible parties continue to search for willing consultants who have technical expertise in weapon design.

It is of some consolation that among the new breed of high-tech terrorists, few have fully mastered the workings of the modern weapons that are or could soon be at their disposal. They place themselves at risk in attempting to assemble, test, and ultimately use them. Breathing viruses, touching blister agents, or handling radioactivity may at first be painless to self-styled weaponeers, but not for long.

Should radical groups succeed in safely assembling a biological, chemical, or nuclear device, or even a cocktail that combines such ingredients with explosives, their device will become an effective weapon only if it releases its destructive power directly onto a target of people or buildings. And then, the carnage will be disappointing unless the perpetrators truly know what they are doing. The effectiveness of even the less-sophisticated pipe bomb, fertilizer bomb, or letter bomb depends on the technical skills of those who construct and set the detonation devices. Hence the scramble to find experienced technical help is on.

Why is a weapons consultant necessary when so much information is readily available elsewhere? Training in the basic principles of advanced weaponry can be self-administered with the help of do-it-yourself handbooks published frequently in the United States. Indeed, addressing the challenge of weapon design is becoming an increasingly popular avocation for Internet users who search for such handbooks and who exchange their latest artistry with one another on the Web.[6]

"Weaponization" of the destructive materials and supporting components is one of the keys. In short, this is the stage when engineers ensure that firing and release mechanisms will function like clockwork, that deadly ingredients will travel in the planned direction, and that flammable materials will ignite at the precise moment to engulf the intended targets. Whether providing a basis for spraying chemical agents, for poisoning water supplies, or for trying for the big bang with a mushroom cloud, the true science of weaponization can only be learned from hands-on trial and error. If the components and materials to be used in weapons are produced or even assembled by novices, experienced hands can help ensure that the pieces are sufficiently well honed so that they will, first of all, fit together and then work in a synchronous fashion.

Experts like Nechai could supply such technical brainpower for this and other critical tasks. When we think of likely candidates for such work, the accusatory finger automatically points to the underemployed veterans of the Soviet military effort as the most dangerous sources of weapon know-how. They may not be the only consultants around, but they certainly are among the most formidable.

There are over twenty thousand specialists throughout the Russian nuclear complex whose knowledge of weapons would be of considerable benefit to rogue states or terrorist groups trying to enter the nuclear weapons era. Some could help bring new entrants into the weapons arena while others have talents better suited for countries trying to upgrade their already limited capabilities. These specialists know how to design, assemble, and test weapons. They are expert in producing the highly enriched uranium and plutonium that provide the explosive power. And they know which approaches will not work under any circumstance, having already been in and out of blind alleys for many years.[7]

Most of them, perhaps fifteen thousand, are located in the ten formerly secret atomic cities whose rundown conditions mirror many of the problems in Snezhinsk. Paychecks are low and erratic, and much of the current activity at the institutes is best described as busy work. Since 1992, the scientists and other workers have been promising (and occasionally delivering) strikes which have prompted the Moscow press to banner such headlines as "The Atomic Cities Are Strike Ready" and "Bomb Designers Threaten To Close the Closed Cities."[8]

With thousands of excess nuclear warheads scattered across the vast expanses of Siberia, neither the nuclear authorities in Moscow nor the Russian people are interested in designing new weapons even if they would be cheaper, more powerful, and more reliable than the earlier designs. Nor are the officials who control the nation's purse strings prepared to pay much to Snezhinsk scientists to monitor the existing weapon stockpiles or to ensure they will really perform as intended, despite the age of the devices. They prefer to believe that the Soviet-era weapons will work if they are ever used. Given the many other competing demands for scarce finances, that seems good enough. Besides, there are enough troubles in the capital that take precedence over worries about people out in the provinces. Out of sight, out of mind is often the outlook in Moscow surrounding inaction on behalf of these institutes.

Alternative high-tech employment opportunities, or indeed any type of income-generating employment openings, are rare in these ten hideaway cities with a total population of 700,000. Moving out of the pockets of isolation is not easy either, given extended family commitments within these cities and the near impossibility of finding apartments elsewhere.

The world would make a serious mistake to ignore this desperate situation. The Chernobyl catastrophe taught the Soviet Union—and the rest of the world—the horrendous consequences that can result when worker discipline collapses in the nuclear industry. In a reorganization move during the early 1980s, the Soviet Ministry for Electrification had assumed responsibility from the highly disciplined Ministry of Atomic Energy for operating the Chernobyl power plant. The new operators, under more relaxed management of the Ministry for Electrification, were simply going to test the limits of the reactor's operating capabilities, so they thought.[9] Many books have been written on the details of the Chernobyl accident, but the significance of the organizational change from a paramilitary ministry has not been given priority in the writings of the historians.

Now, in the atomic cities that were founded in the most extreme paramilitary atmosphere, discipline is gradually collapsing, not because of reorganization but due to resignation to new economic realities. Underemployed workers have few incentives to continue to diligently serve the state. They may not be paid at any given time. Drinking is on the rise, and theft and street crime have penetrated the high fences. Schemes that border on illegality are slowly becoming methods of choice to compensate for personal hardships resulting from reduced subsidies. With secrets being the most valuable commodity, the situation poses a serious threat. Beyond the ten atomic cities are another fifty or more formerly closed cities where other nuclear, chemical, biological, missile, aircraft, and laser technologies, together with associated brainpower, are in a dangerous limbo.[10]

Meanwhile, the U.S. government may have inadvertently encouraged the diffusion of weapons technologies from Russia. A few examples will illustrate the point.

Six months before Nechai's suicide, another weapons pioneer from the Urals region had lost his life, this one by assassination in Yekaterinburg. Vladimir Smirnov was a missile engineer. Heavily decorated by the state, he had designed the sophisticated Soviet missile system known as SV-300V. This weapon system has been described as the Soviet counterpart to the American Patriot missile and also as the Soviet defense against American aircraft cloaked in stealth antidetection technology.

Smirnov had not been content to sit back and suffer the loss of his income while other self-styled Russian businessmen systematically diverted Russian military assets for their own personal financial gains. Responsible for the employees at his manufacturing plant, as well as for his own family, he elected to help arrange the sale of a SV-300V missile system to American intelligence agencies. The price reportedly was $60 million.

In March 1996, unidentified perpetrators murdered Smirnov. The reasons for the murder remain shrouded in a cloak of obscurity despite high-level interests at the time in both Moscow and Washington. Though a large team of Russian investigators was dispatched from Moscow to the scene, their report of the incident will probably never be released. Speculation abounds that either Russian nationalists were unhappy with Smirnov's audacity to sell a technological jewel to the Americans or, more likely, the Russian mafia, which thought it would receive a substantial cut on such sales, felt betrayed because its share was much smaller.[11]

The techniques employed by the American agencies in obtaining the missile are held under tight security wraps.[12] Perhaps American agents spotted the missile, together with the designer, at the biannual arms bazaar in Abu Dabai, United Arab Emirates, where Russian entrepreneurs frequently put their technological assets on the auction block. From the Western perspective, it was better for the Americans to be paying dollars for this sophisticated weapon than for operatives from the Middle East to be providing baksheesh for the technology.[13] Still, the transaction underscored that many Russian high tech items are for sale if the price is right—not too high, but right. After the SV-300V sale to the former enemy of the country, Russian enterprise directors argue they are free to sell their technologies to anybody as they seek buyers in countries on international blacklists.[14]

Another unusual story about the aggressiveness of U.S. agencies in encouraging the packaging of weapons information for use abroad came to light in late 1996. An intrepid American reporter learned that several years earlier the Pentagon had purchased a large array of technical documents concerning the testing of Soviet nuclear warheads. For less than $300 thousand, a leading physicist from Sarov, the other Russian nuclear weapons design center in addition to Snezhinsk, had agreed to enlist more than two hundred of his colleagues in writing reports for the U.S. Defense Nuclear Agency (now named the Defense Special Weapons Agency) on their contributions to nuclear testing. For $500 each, they were to share their insights concerning 715 Soviet nuclear tests. They apparently did so, with the product being a weighty two thousand-page document. Undoubtedly, they thought they were taking the first step to becoming comrades-in-arms with a former adversary. Several expressed disappointment to me when there was no demand for the second volume, at least from Washington. Apparently, all the cream was skimmed on the first deal.[15]

At the time the contract was initiated, the Pentagon refused my request for information on the undertaking. Along with Russian colleagues, I heard about the project through the Russian rumor mill. We were concerned that such intelligence gathering could jeopardize other American efforts to engage the same Russian specialists in less sensitive research activities with longer-term benefits. In response to my interest in reviewing copies of the reports, Department of Defense officials explained that once they received the Russians' manuscripts, their reports could not be distributed. This procedure apparently was intended to prevent proliferation of the information. While such apprehensions were valid, it wasn't long after the documents were completed that some Russian authors were reported to be offering similar monographs to other buyers from unknown foreign institutions.[16]

After the press broke the story, the Russian government reviewed the roles of several Russian senior scientists in organizing this sale of sensitive information about nuclear testing for the equivalent of a few truckloads of vodka. While other Russian scientists initially bristled over the idea of colleagues selling former secrets for money, the critics soon became more concerned about the stupidity of their fellow scientists in accepting such a low price. They reasoned that their colleagues were simply trying to survive after the state pulled the rug from under them. To them the issue wasn't that someone had sold sensitive information, but that they had negotiated such a poor price for it.[17] The larger lesson was that security barriers were being lowered in allowing data to flow from the heart of the Russian nuclear security complex to the United States and perhaps to other buyers as well.

A third endeavor of the Pentagon also had near-term payoff for American intelligence agencies. To some, it legitimated the sharing of information previously considered sensitive. To others, the low prices dampened enthusiasm for further cooperation with the United States in areas related to military research. This story also became well known throughout Russia in 1996.

For decades, the Soviet Ministry of Defense closely guarded an information center in Moscow which served as a central repository for documents on military technologies. Like all institutions in Russia, the center suffered drastic budget cuts in the early 1990s. Again, paychecks were delayed. Soon American defense contractors were knocking at the center's door, ready with cash to purchase any documents that could be released. The center quickly sensed an opportunity for a financial bonanza. In the midst of administrative turmoil in Moscow as to who had authority to release documents, many previously classified packets were sold at fire sale prices to American purchasers, who could have paid more. In 1996, when details became known to the Russian government, opportunities for bilateral cooperation in sensitive areas suddenly were more constrained than might otherwise have been the case. Key politicians apparently were divided. Some wanted to protect the technology that had always been the Soviet

8. Two suicide bombers blew themselves up in a crowded outdoor market in Jerusalem on July 30, 1997, killing at least 13 and wounding about 150 others.

trump card while others wanted to know whether better prices could be obtained from other countries.[18]

Rogue states are clearly interested in designs of nuclear weapons systems, blueprints of missile components, technical data about the functioning of advanced weapons, and details of many less sophisticated technologies. We are aware that terrorist organizations have close ties to rogue governments, particularly in the Middle East. The stronger the capabilities of these states that target technical acquisition efforts on Russia, the more dangerous will be the associated terrorists within and outside the governments. Also, if Russia is willing to part with information on nuclear testing and antimissile systems to the United States or anyone else, then foreign shoppers looking for radioactive wastes to disperse in populated areas or for new versions of bazookas should not have difficulty finding sales outlets for these less sensitive items.

The plummeting economic conditions at Snezhinsk are not unique among the dozens of cities which birthed some of the most destructive weapons of the twentieth century. If a major concern is that of the transfer into the wrong hands of formulas for designing armaments, the situation there is not the most worrisome. Access to Snezhinsk is difficult. It is many miles off the beaten track; multiple layers of barbed wire fences still impede access; and the work force is not experienced in selling its wares in unfamiliar marketplaces. Still requiring special permission to enter, Snezhinsk simply is not a visitor-friendly place.

In sharp contrast, many lightly guarded institutes with large numbers of disenchanted weapons specialists abound in and near Moscow, St. Petersburg, and Nizhny Novgorod. Immediately following the dissolution of the Soviet Union, these cities became high-tech bazaars waiting for the highest bidders. Foreign traders of all stripes operate low-profile enterprises behind unmarked doors in buildings throughout these cities. Many past Russian inhibitions on deal-making have given way to aggressive searches for high-stakes transactions.

Whether considering conditions in Snezhinsk, in other remote cities of the former Soviet nuclear complex, or in the major metropolitan areas of Russia, nuclear and other types of weapons expertise have surely been leaking out of the country. The key questions are, "How much is leaking?" "How significant is the information?" and "To which destinations is it headed?" The following examples only touch the leakage potential.

- In 1993 in Moscow, I met with two confident Russian missile designers from a group of twenty-five who had attempted to take their brainpower to North Korea. They had only recently been intercepted at the Moscow airport, en route to Pyonyang, by Russian security forces. At the last minute, the Russian Foreign Ministry apparently had second thoughts about the international ramifications of sending missile experts to North Korea allegedly to help in designing a communications satellite system. Official permission to travel was promptly reversed, and the engineers returned to their institute in the southern Urals where they began to search for other sources of financial support. Following our consultations as to American interests in the technical wherewithal of the institute, they succeeded in engaging an American firm in a joint venture to build gyroscopes for civilian aircraft.[19]

- Russian specialists from the Moscow Institute of Graphite described to me their plans for helping Indian colleagues improve the heat-resistant properties of rocket engines. They could mold graphite with a degree of precision unmatched in the world. Again, however, Russian officials caved in to international pressure to abstain from such questionable dealings if Russia wished to retain its status as a responsible trading partner with the West. When the diplomatic brakes were applied, the institute redirected its attention to research on fullerenes, a somewhat exotic material with possible, future weapons applications.[20]

- In the town of Zelenograd on the outskirts of Moscow, I encountered electronic specialists who had found new outlets for their expertise—in Syria. Steady incomes could be assured through long-term employment contracts in that country. While joint ventures in Zelenograd with American and Korean electronics giants were important, these companies could accommodate only a few of the former Soviet military researchers who were out of work. Also, from the technical aspect, most of the jobs were not very challenging. The Western companies usually limited Russian contributions to the more routine activities rather than entrusting Russian specialists with cutting edge technologies. In Syria, where electronic systems for supporting either military or civilian work were in an embryonic state, there were no such inhibitions on innovations that could save money or improve performance.[21]

• As already suggested, the biggest market for Russian high-tech expertise has been China. This country has been eager to subsidize scientific exchange visits and to purchase Russian equipment. The Chinese seem to be interested in all types of Russian technologies, from the design of rocket engines to techniques for bioengineering of microbes. Such interest is rooted in the extensive scientific and industrial interactions between China and the former USSR during the 1950s and 1960s. Political problems interrupted such interchanges for nearly two decades.

In 1994, I encountered Chinese specialists in residence at Snezhinsk who professed their sole interest to be peaceful nuclear research. I learned they were negotiating purchases of hydraulically powered hospital beds from this weapons center. Was such a transaction a teaser to set the stage for cooperation in more sensitive areas? My Russian colleagues gave less attention to the possibility of covert Chinese agendas than to the reality of overt assembly of beds for immediate cash.[22]

For many Russian researchers, travel to China has become quite common. In Novosibirsk, for example, Russian physicists explained to me how they had developed a cash cow in China, initially through sales of nuclear accelerators to that country. By 1994, they had sold seven units at a total price of over $5 million. They didn't know—or care—how the Chinese put these accelerators to use, although they pointed out that in the Soviet Union such high precision "heat generators" had been used for annealing metallic wiring to plastic insulation in fighter aircraft.[23]

In summary, Russian institute directors have inherited large storehouses of technical information and warehouses of scientists. Their information and expertise would be of considerable potential value for governments and terrorist organizations seeking shortcuts in building up their advanced weapons capabilities. Soviet trademarks on high-tech items have already found their way into many developing countries that have been breeding grounds for militant groups (e.g. Libya, Iraq, Iran, Syria), and the proliferation of weapons technologies of previous years will undoubtedly continue as scientists and engineers of all stripes circulate into and out of Russia.

Neither should we discount the possibility that truly desperate people are lured by organized crime syndicates that have emerged in recent years on these cities' fringes. Such powerful contacts would surely have some ideas as to how skilled scientists could be of value to connections in the Middle East and other areas; they would be more than happy to broker headhunting transactions. As the Smirnov murder probably demonstrates, these contacts want an appropriate cut of the action when profitable sales are in the offing.

This is not to say the Russian government and the international community have not tried to dampen the unrestrained pursuits of the institutes and enterprises to find customers of any stripe. The Russian government has pressed hard for their own conversion programs that focus on peaceful applications of advanced technologies, but without funds these are hollow promises. Meanwhile, the security services attempt to ensure that national secrets remain in Russia, but the guidelines defining secrets have turned out to be more elastic at times than previously envisioned.

When economic problems radiated throughout the newly independent states of the former Soviet Union at the end of 1991, the United States and other Western countries immediately responded with airlifts and sealifts of humanitarian aid to help the populations through their initial difficult days. At that time, a friend from one of the atomic cities pleaded with American and European officials to consider humanitarian aid for the ten closed cities. He predicted that unless there was overseas relief, the falling standard of living would soon raise major national security problems both in containing the expertise in the cities and in protecting the dangerous nuclear material housed in the cities.[24]

While my Russian colleague acknowledged the scientists had been spoiled by their former subsidies, he correctly warned that ignoring these cities could turn the most loyal scientists of the Soviet era into disaffected groups prone to accidents and even to misuse of government property. His pleas went unheeded. The weapons scientists had been the privileged class; they did not deserve humanitarian aid, or indeed any other type of aid, argued the American and European foreign assistance experts. Besides, they added, the West could not be perceived as supporting a "military" city.

Fortunately, Senators Sam Nunn and Richard Lugar, together with Secretary of State James Baker, recognized that an unstable population living close to lethal weapons could add new definition to the metaphorical hungry Russian bear—one now willing to trade armaments for a few jars of honey. They took the initiative to devise not a foreign aid program but a program of cooperation that would serve U.S. security interests as well as help the Russian scientists find new professions. Thus, the Western countries, led by the United States, have initiated a number of programs conducted under the auspices of the International Science and Technology Center in Moscow and other organizations to provide opportunities for some former Soviet weaponeers to work on civilian projects. This approach should reduce economic incentives for them to transfer their know-how to states and groups of concern. The success of these programs has exceeded the most optimistic predictions of the early 1990s, with more than 10,000 former nuclear weapons specialists—and a comparable number of aerospace, chemical, and biological scientists—involved in civilian activities from 1994 to 1998 under international oversight. While there are still tens of thousands of other weapons specialists struggling to survive, an impressive start toward career changes has been made.[25]

Most other foreign programs targeted on former Soviet weaponeers have come about very late, and a great deal of Soviet technology is already being disseminated to many corners of the world. It is true that, contrary to early predictions by Western experts, emigration from Russia of weapons specialists to countries where they could interact with groups seeking advanced weaponry has been very limited. Nonetheless, more worrisome as emphasized before are the thousands of specialists on the loose in Russia itself, including some who may be willing to sell their knowledge of weapons systems to the highest bidders who pass through the country. A few probably have relevant documents in their possession or can obtain or develop such documents. Some have ready access to the Internet.

This does not mean that a foreign agent, even one with un-limited financial resources, can simply go to Russia and immedi-ately obtain the necessary documentation and knowledge to design and build sophisticated weapons. What it does mean is that well-financed teams of partially trained weapons-design sci-entists and engineers from developing countries may be able to obtain from former Soviet weaponeers specialized information that will reduce the time and cost for developing and constructing advanced weaponry for hostile endeavors.

What additional steps can be taken to interrupt this outward flow of technical data and hands-on experience? Unfortunately, very few in the short run. While arms control and export control agreements will limit diffusion of military hardware and may even provide for inspection to help ensure compliance, agree-ments to control the transfer of knowledge are far more difficult to devise and to police. Export control procedures, for example, are intended to control flows of sensitive technical data as well as equipment, but this is an almost impossible job, particularly when the information is stored in people's heads.

Until Russia is back on its feet economically and the corrup-tion that now permeates many governmental organizations goes into remission, the powder keg will continue to hover over the flames. Organized crime will win battles over who owns store-houses of technology—both equipment and technical data—that resulted from the trillions of dollars of Soviet investments in mili-tary systems dating back to the end of World War II. This possi-bility alone should be enough to bolster support for continued U.S. funding of cooperative projects that work to limit the outflow of sensitive commodities and technologies. In many ways, the American-initiated counter-proliferation programs have been the only effective shows in town, and their continuation is a critical stop-gap measure.[26]

Four incidents highlight the scope of the problems of tech-nologies spreading in illicit ways.

(1) *Of great concern are the restless designers of nuclear weapons stung by the 1996 international ban on testing of nuclear devices.* No longer can they attest with full confidence to whether their weapons designs really work, and gone is the opportunity to check improvements on the original plans. As noted previously, in frustration and perhaps uncertain of future job security, should even a single underemployed weapons designer entertain an offer from a militant group, such misplaced expertise could represent considerable vulnerability here or in some other part of the world.

One dimension of this technological deprivation was driven home to me in late 1995 when the Department of State invited me to a meeting of American and Russian nuclear weapons designers in Washington. The agenda would include presentations on the use of high-speed computers, developed for weapons programs, to create models of environmental problems and related issues, so I had been told. On the very day of the meeting, U.S. and Russian diplomats in Geneva announced agreement to ban all types of underground nuclear tests. Registering their protest in Washington, these bomb designers made a heated presentation claiming that a test ban was a bad idea. A ban would prevent them from periodically checking to ensure that the existing weapons worked, thereby reducing the deterrent value of the weapons.

Following the discussion, several U.S. officials expressed to me their concerns about the ambitions of the informal international network of nuclear bomb designers that seemed to be emerging. "Will the American specialists encourage Chinese counterparts to continue testing?" was the most penetrating question after the meeting. Weapons designers all over the world passionately and correctly believe their efforts have been essential to global stability, and some are convinced that a limited number of countries should now retain the right to conduct tests. They simply don't agree that a comprehensive test ban is an essential step if we are to successfully advocate no testing in rogue states and thereby prevent the evolution of a world of high-tech anarchy. However, they offer no alternative approaches as we try to build firewalls between nuclear weapons and the terrorists who aspire to use them.

As a footnote to the meeting, rumors were rampant in Washington and in Beijing shortly thereafter that the Chinese were opposing an international ban on nuclear weapon testing. Reportedly, they were following advice received from Russian weapon designers over the importance of keeping open the option to test.

(2) *Russia is not the only repository of Soviet weapons expertise.* By coincidence, I visited Tbilisi, Georgia, in 1995 three days after the visit of the former president of Iran, only to learn that the president's entourage had scheduled a series of interviews with nuclear scientists from Georgia to entice them to move to Iran. At the time, the average salary of the scientists was less than $4 per month, but they still resisted. Few, if any, Georgians are interested in living in Muslim countries with unfamiliar lifestyles and working on projects opposed by the West, regardless of their economic plight. When I returned to Georgia in 1996, I learned that not a single interviewee had moved to the Middle East, and that their spirits had risen considerably as their average salaries soared to $14 per month. I also learned the Iranians were still trying to entice some physicists to Teheran.[27]

(3) *Recruitment of expertise may come from unusual venues:* In 1993, I discovered an advertisement of the Hong Kong Sun Shine Industrial Company circulating in Russia. I later learned during congressional hearings this company was well known to American intelligence agencies for its trading in conventional armaments.[28] The advertisement suggested that high technology may have become a busy recruitment area for international soldiers of fortune.

> We sell gun, machine gun, ammunition, rocket gun, rocket launcher (projector), mortar, tank, amphibious tank, armored carrier, landing ship and boat, and other military supplies which are made in China. Especially we sell middle and short distance guided missiles like, M-8, M-9, M-10, and M-11. On [the] other hand, we have detailed files of hundreds of former Soviet Union experts in field of rocket, missiles, and nuclear weapons. These weapon experts are willing to work in [a] country which needs their skills and can offer a reasonable pay.[29]

(4) *Money attracts eager bedfellows*: Bruno Stemmler claims he went to Iraq in search of contracts for developing irrigation systems and new energy sources. Could a German engineer who had

spent a lifetime designing and testing gas centrifuges to produce enriched uranium really believe the Iraqis would welcome him in Baghdad because of his knowledge of irrigation systems? Should we believe his contention of being totally surprised when, in his first meeting in Baghdad, the principal subject was the design of centrifuges?

During the late 1980s, Stemmler became unhappy with his managers at the Munich company that built centrifuges for producing uranium for nuclear reactors. He disputed his paycheck size and complained of the general lack of recognition of his support for the company's efforts. The breaking point apparently came when he sought 50,000 marks for his contribution to advanced technology patents. When the company offered 15,000, he left in protest.[30]

In 1988 and again in 1989, he purportedly paid secret visits to Iraq to assist nuclear specialists in designing centrifuges for enriching natural uranium to the level used in nuclear weapons. He advised on desirable specifications of ball bearings, vacuum lines, and cascade systems. Additional scheduled visits were interrupted when German security authorities seized him at his home in Munich, accusing him of giving away state secrets. He was never tried for a criminal act. Until the end, Stemmler steadfastly contended he did not discuss secret information nor violate any laws. Since then, Germany has enacted export control laws that presumably forbid such activities. However, the adverse publicity surrounding his involvement in the Iraqi nuclear weapons program prevented him from finding new employment in Germany.[31]

The Iraqis were presumably receiving advice not only from their own technicians but also from other foreign specialists at about the same time Stemmler was visiting Baghdad. Thus, his recommendations were undoubtedly considered within the context of other technical views. Stemmler's extensive experience in designing centrifuges offered the Iraqis an important technical perspective which was well worth the $75 thousand paid for his services.

Meanwhile, we need only look in our own backyard in the United States for training grounds for scientists and engineers the world over. From the far-reaching laboratories and training facilities of government to the hallowed halls of academia to the conference tables in corporate boardrooms, the seeds of weapons expertise are planted to be nurtured for the defense of the United States. Unfortunately, they are sometimes harvested by those with other less noble plans.

Stepping back from nuclear expertise to more conventional weapons, for decades the industrialized countries, with the United States leading the pack, have provided their developing neighbors with all manner of conventional armaments and the training packages to go with them. We provide weapons valued at billions of dollars through sales and subsidies each year. During the Persian Gulf War era, for example, the United States transferred more than $17 billion of military hardware to Middle East countries from 1992–1994. Saudi Arabia received 60 percent, Egypt 12 percent, Israel 9 percent, Kuwait 6 percent, and *Iran* 5 percent. Such a surge in the region was understandable. Looking not only to the Middle East but also to Asia and Latin America, what has happened to equipment transferred with approval of the Department of Defense? And no less important, what has happened to the personnel trained to use the equipment?[32]

From West Point, where I studied with classmates from the Mexican and the Philippine armies, to airbases in Texas, where pilots from all over the world practice flying in combat missions, many foreign military trainees have learned their trades at the expense of American taxpayers. From 1950 to 1995, the Pentagon trained over five hundred thousand foreign personnel, almost all from developing countries, under the International Military Education and Training Program. Prior to 1980, more than four hundred trainees came from Iraq, eleven thousand from Iran, and five hundred from Libya, for example.[33] Again I ask, where are these trainees two decades later? Perhaps some of the Iranians are in jail, while surely some remain career soldiers in that country and in other countries. The real question is: "Are some seeking to apply their skills in dangerous ways?"

No longer emphasizing technical subjects, the Pentagon's cooperative programs now address benign topics such as accounting procedures associated with contracts, licensing and royalty arrangements, opportunities for joint ventures in developing military products, and organization of bilateral and multilateral cooperative programs. Of course, sales of military hardware still include hands-on technical training, and the United States continues to be the world's largest arms merchant. The most sophisticated items usually go to European allies, although highly advanced technologies were certainly available to Saudi Arabia during the Gulf War.[34]

In short, arms sales, particularly conventional weapons such as small rocket launchers and machine guns, and associated training for today's soldiers are key elements in addressing the risk of global terrorism; such technologies can combat terrorism, but in the wrong hands they can contribute to achieving terrorist aims. In 1995 President Clinton adopted a Conventional Arms Transfer policy. The policy warns of the dangers of proliferation of military technologies to rogue states through sales and other types of transfers in regions of concern. It provides criteria to be used in decision-making on transfers of U.S. arms. It is a starting point but needs continuing refinement as terrorist plots encroach further and further into activities that previously had been considered legitimate programs to defend rights of national sovereignty.[35]

The Pentagon is not the only venue for military-related training. For decades, America's universities and technical colleges have proved exceptional places for initial training of future weaponeers from not only the United States but from nations throughout the world. Basic physics, chemistry, and biology concepts, the grist of all universities, underlie weaponry. Technical design and system integration courses, the starting points in weaponization programs, have become standard fare for engineering students.

While educational opportunities with some relevance to weapons systems are offered at universities in many countries, several characteristics set apart an American education from other educational programs for aspiring high-tech terrorists. First, many American military specialists are associated with university programs as adjunct faculty, visiting lecturers, participants in symposia, and even part-time students. Second, Massachusetts Institute of Technology, Stanford University, and many other American universities take pride in their close working relationships with American industry (including many companies from the military–industrial complex) with students frequently being offered internships to work on industrial projects at the universities or at company facilities. Also, the U.S. Department of Defense sponsors billions of dollars of research activities at American universities, some classified but most unclassified, with little attempt to dictate the selection of students to work on unclassified projects. Finally, the high-tech database resources available to students at American universities are unparalleled in the world.

In the mid-1970s, India's nuclear testing raised a red flag over the education of foreign students in the United States in nuclear-related subjects—education that might directly enhance the nuclear weapons capabilities of not only India but also of Iran, Pakistan, and other countries. At that time, the U.S. government concluded there were limited risks that American-based education would enhance weapons capabilities of developing countries since university programs did not involve access to classified information or hands-on weapons training. However, in the early 1980s, our government took a second look and began denying visa applications of students in nuclear science and engineering from some countries that were of particular concern.[36]

A decade later, revelations that Iraq had sent large numbers of students abroad, primarily to Europe, for nuclear training revived interest in more stringent restrictions in the United States related not only to nuclear science but also to chemical and biological technologies. The topic of foreign student training remains controversial in Washington given the technical and cultural benefits in allowing students from the developing countries to obtain the technical skills necessary to promote economic development in their countries. It surely is not always possible to distinguish between students with praise-worthy goals and those with suspicious motivations.[37]

At the same time, a consistent policy with European countries seems essential. If we cannot convince some of our traditional allies to show restraint in sales of sensitive items to rogue states, we undoubtedly will have difficulty convincing them of dangers associated with exchange students.

Beginning many years ago, no country has been more aggressive than China in taking advantage of science and engineering training opportunities in the United States. As early as the 1950s, future leaders of China's nuclear and missile programs were refining their skills at American universities, both as students and as faculty members. Qian Weichin left the California Institute of Technology in the 1950s to organize the Chinese missile development program. At least four American-trained nuclear scientists—Jiang Shengjie, Chen Nengkuan, Deng Jiaxian, and Guo Yonghuai—had leadership roles in the early days of China's bomb development program. In the 1950s and 1960s, China's close ties with the Soviet Union meant a steady stream of personnel exchanges between Peking and Moscow related to nuclear programs but American training played an important role as well.[38]

In the 1970s and 1980s with the opening of China, the influx of tens of thousands of state-sponsored Chinese science and engineering students into the United States was overwhelming. Almost all enrolled in science and engineering courses at leading American universities, establishing a huge presence in American research laboratories. To American enthusiasts for the programs, including the American professors who acquired highly talented, low-cost assistants, the influx was described as leading to political conversion of an elite group of Chinese leaders of the future to Western ways of thinking and acting. Critics of the influx predicted a drain of sophisticated American technology flowing to China for many purposes. Both the enthusiasts and the critics have been proven correct.

For decades American companies have been targets of visitors seeking to uncover industrial secrets. Many techniques are widely accepted practices: acquisition of patent information; purchase and testing of new products with a view to reverse engineering of comparable designs; and collection of all available documents at industrial trade fairs and symposia for subsequent sorting, reproduction, and distribution. Also, unclassified U.S. military databases and information centers have become standard stopping points for technology seekers.[39]

That said, it is important to remember that no one knows the extent of foreign cherry picking of American technological knowledge. The distinctions between legal and illegal document acquisition have become blurred, and restrictions on the export of technical data have lost much of their meaning with the advent of new electronic systems for transferring databases.[40]

As I noted earlier, large numbers of dual-use technologies of potential interest to terrorists have applications in both civilian and military systems. For example, a sophisticated nuclear weapon has over four thousand components in addition to the "physics package" containing the nuclear material. A simple device needs many hundreds of such components, for providing power, arming and triggering the device, and ensuring synchronization of the interacting parts. Most of the components are identical or very similar to parts used for civilian applications: springs, spacers, connectors, switches, relays, condensers, wiring, and many other nondescript items for mechanical and electrical systems.[41] Hence, it may be impossible to determine whether the purchaser is obtaining inventory for perfectly harmless products or stockpiling supplies for destroying civilian structures.

Many new dual-use products are now penetrating markets in regions of the world where simple technologies had previously been the rule. Exemplifying the problem, electronic systems to guide military forces may be identical to electronic systems for civilian applications. Materials that resist high temperatures— new plastics and metals, for example—may find both civilian and military uses. Two examples seem to tell it all. Titanium alloys were developed for military aerospace applications but are now widely used in golf clubs and golf balls. Computers that can calculate missile trajectories can also be used for determining bank balances.

The multinational companies, whether rooted in the United States or in other countries, are important conduits for the dissemination of dual-use technologies. After all, they developed the technologies that are now their lifeblood. Some particularly sensitive dual-use technologies are subject to export controls designed to limit their dissemination to those with less than reliable reputations. The export groundrules are often uncertain, and mechanisms for enforcing export limitations are less than perfect. Even the United States, with the best export control system in the world, has difficulty keeping rein on all technologies of concern.[42]

In the years ahead, American firms marketing high-tech products will be in fierce competition for foreign commercial markets. This competition will inevitably lead to the introduction of advanced design and manufacturing capabilities in regions inhabited by international terrorist groups. Despite the care and prudence exhibited by American companies, or by their competitors, as high technology spreads so does the enhanced world-wide capability to produce ever more destructive weapons.

Against this somewhat disheartening background, it is encouraging to learn of the highly effective visa-screening program of the Department of State to prevent known terrorists from entering the United States. When such aliens are allowed in to attend U.N. meetings or for other overriding reasons, our security agencies are alerted to their presence so they can watch their movements.

The State Department now has more than 35,000 marked foreigners in its sights. (See Table 11.) For any one of them to receive an entry visa is a precarious proposition. Of course, the department's list of known terrorists is not 100 percent complete, it may not include "first-time" terrorists, and the screening process is not infallible. The program does draw on all relevant information in Washington, terrorists with eyes on American targets have been put on notice that their actions will be observed.[43]

In responding to the terrorist threat against the United States, today and tomorrow, our diplomats, intelligence services, and

TABLE 11.   Visa Tipoff List of Terrorists and Other Felons

No. of Records	Category
3454	Intelligence Agent
145	Representative of Foreign Terrorist Organization
2061	Member of Foreign Terrorist Organization
2464	Has Engaged in Terrorism
6988	Likely To Engage in Terrorism if Enters U.S.
390	Hijacker
658	Hostage Taker/Holder
28	Attacker of Internationally Protected Person
1464	Assassin
558	User of Nuclear/Biological/Chemical Agent
2532	User of Explosives or Firearms
4135	Prepares/Plans Terrorist Activity
449	Gathers Intelligence on Potential Targets
5844	Provides Support: Safehouse, Weapons, Money, Identification, etc.
155	Solicitation of Funds for Terrorist Activity
4932	Solicits Membership or Terrorist Acts by Others
675	Lost/Stolen Passports in Terrorist Hands
120	Silent Hit
1095	Additional Intelligence Required
38,147	TOTAL RECORDS

Source: Department of State, July 1996

law enforcement agencies must give priority to curbing the activities of those individuals who carry dangerous technical expertise, whether it be nuclear technologies or the techniques for packing and detonating plastic explosives. We certainly can agree that cooperative programs to redirect efforts of former weapons scientists to peaceful pursuits are important, particularly in Russia, and the procedures that have been developed to screen applicants for American visas are essential. How can we address the possibility that visitors to the United States, in the guise of students or military or industrial trainees, may have hostile intentions in mind? What additional steps can be taken to maintain the distance between dual-use technologies and the hands of terrorists?

Military and industrial training projects can be carefully orchestrated. However, controlling the flow of technical information within U.S. universities is neither desirable nor feasible. If aspiring terrorists are denied entry to American universities, they will find other schools. Therefore, we must live with the reality that weapons-related information will be available globally. It is when the knowledge is used to construct or use weapons that we have the best chance of disrupting terrorist schemes.

That said, it still makes little sense to allow fanatic Iranians who are known to be bent on harming America to study at our universities as indicated at the outset of this chapter.

Drug Runners and Terrorists: New Partners in Crime

Drugs account for 2 percent of the global economy.
The International Monetary Fund, 1997

*The combination of the terrorists and the drug traffickers
is a very tough thing, because the drug traffickers have
such immense sums of money and the terrorists provide
them with the muscle. It's quite a combination.*
Former Secretary of State George Shultz, 1991

*The terrorist group FARC has financed its activities
with $600 million from the Colombian drug dealers.*
Colombian President Ernesto Samper, 1997

*I*n January 1996, just south of Monterrey, Mexico, fifteen Mexican drug agents in minivans smashed through the front gate of a hideaway of Garcia Abrego. After a short chase, they seized the Mexican drug dealer by the shirt as he tried to hoist his portly frame over a back fence. Abrego was accused of shipping cocaine. Indeed, he had shipped up to one-third of the cocaine consumed in the United States during the previous decade. Abrego was not worried. He joked with his captors that they would be hearing from him soon. He was confident the Mexican system of criminal justice would not contain him for very long. Needless to say, he was in a state of shock as he struggled with the drug enforcement agents who summarily dragged him aboard a jet airplane destined for the United States where he didn't "own" the law enforcement machinery.[1]

As the first international drug dealer included on the FBI's most wanted list, Abrego was a prize trophy the Mexican president decided was too hot to handle. Abrego served the Colombian cartels, distributing their cocaine in six American cities, and his personal income approached $2 billion per year. The millions of dollars in bribes he had systematically dispatched to officials from former Mexican presidents on down were simply petty cash.[2]

In July 1997, *Newsweek* reported that another Mexican kingpin, Amado Carrillo Fuentes, who had an estimated drug fortune of $25 billion, was "liposuctioned" to death in Mexico. After undergoing eight hours of plastic surgery for a facelift that resulted in a stump of a nose, eyelids scarred by surgical incisions, and a chin reshaped by implants, a person reported to be Fuentes died of heart failure. "Liposuction" may not be the technically correct description, but clearly the surgery was extensive. American officials confirmed the identity of the corpse through fingerprint matches, and Mexican police claimed DNA matches. Some American law enforcement skeptics thought the death setting to be simply too convenient for someone who at one time had the country's antidrug agency in his pocket and could buy a small country if necessary. While only Fuentes' mother knows for sure whether the body was of her son or of a surrogate, most law enforcement experts believe he is out of the drug business for good. To add to the intrigue, the bodies of two of the surgeons who had performed the operation were found partially encased in concrete blocks several months later.[3]

Fuentes' unique management style will not die so easily. Developed at his Mexican headquarters within eyesight of the intelligence center of the U.S. Drug Enforcement Administration (DEA) in El Paso, it was to: remain invisible, bribe the federal police to serve as bodyguards, kill all participants in operations that go afoul, and sleep only with North American beauty queens.[4]

Earlier, during the early 1980s, the Medellin and Cali drug cartels, the principal Colombian distributors of cocaine, had also suffered crushing defeats. The kingpins of the cartels had been hunted down by the Colombian government not only as drug dealers but as supporters of narcoguerrillas (a term used to define terrorists who take a slice of the drug profits in return for promoting violent intimidation of government officials and multinational companies). Sometimes referred to as narcoterrorists, they have a symbiotic relationship with the drug lords and their kingdoms, joining forces to threaten lives in the cities and the countryside.[5] Reports of the demise of these cartels raised expectations in the United States as well as in Colombia. The hope was the local authorities would not only begin to reduce drug processing but would also stop the brutality that intimidated politicians, businessmen, and farmers. However, the hundreds of groups that have spun off from the major cartels and the independent-minded splinter groups had their own ideas. Many learned from others' errors and changed their style of operation. They established new communication links that allow the kingpins to direct their business dealings through local surrogates, even from abroad. In short, drug czars are increasingly calling the shots by remote control.[6]

High-visibility vignettes depicting the fall of drug lords will continue to enrage, amuse, or confuse the public on both sides of the border. Governments will widely advertise their success whenever a high-profile figure is brought to justice. But, law enforcement agents know that behind each fallen leader is a protege who will promptly step forward to take over the reins. The system doesn't die; only the players change. We can only hope the replacements will be somewhat deterred, at least temporarily, by the state of their predecessors.

The profit margins are tremendous: the initial cost to the traffickers is about $20 for the coca leaves required for one kilogram of cocaine while the price on the street of the drug in 1997 was more than $20 thousand a kilogram. After deducting all expenses, the return is estimated at several thousand dollars per kilogram. Many overseers of such a business can live handsomely on profits.[7] With such incentives, there's no secret as to why business is booming.

Only a small fraction of the money goes to the farmers who grow the crops. Most of the balance of the mega-markups goes to sophisticated crooks, the criminal organizations and individuals who process and distribute the drugs and handle the payments. But, make no mistake, a substantial cut goes to terrorist groups, allied in various ways with the criminals, who have killed thousands of victims standing in the way of their schemes.[8]

Our most immediate concern when it comes to the flow of narcotics into the United States is the destruction of tens of thousands,

9. A farmer checks out his upcoming harvest of poppies in a field in Pakistan.

and possibly hundreds of thousands, of young American lives every year. Almost all of the cocaine and heroin on the streets of America comes from abroad, and 65 to 75 percent of the marijuana is imported. In early 1998, statistics indicated there were 13 million users of narcotics in the United States, including 3.6 million who were compulsively addicted to drugs. Of them, 500 thousand were on heroin, while more than two million crack addicts use heroin to cushion the "crash" that follows the euphoria after using crack. This growth is attributed in part to the high purity (95 percent) of Colombian heroin that can be snorted like cocaine, thus reducing the fear of contracting AIDS from dirty needles. Sadly, many young people are included in these numbers. Starting in the middle school years, one out of five school children uses drugs.[9]

Going beyond drug addiction, it is becoming ever clearer that the vast stores of money generated by the international drug trade have become magnets for those in need of resources to support their terrorist aspirations. For example, in Latin America narco-trafficking has corrupted governments and businesses while nurturing, financing, and arming violent groups that in turn further threaten and intimidate governments and claim many innocent victims. In Peru, the Sendero Luminoso (Shining Path) terrorists extort levies from growers and transporters of coca while accepting money from the drug dealers to protect them from the army and police. Dedicated to a peasant armed struggle to establish a leftist, ethnic Indian state, this group has killed over twenty-five thousand people in recent years, often in bloody public executions featuring bodily mutilations. Terrorist groups in Colombia have worked hand-in-glove with drug lords to challenge the political and legal structures at all levels. Complicity in the drug trade by violent groups that are clearly terrorist in their behavior has also been uncovered in El Salvador, Mexico, and Brazil.[10]

In other areas of the world, dissident elements in Afghanistan and Pakistan rely on drug money to sustain their terrorist tactics aimed at changing the political structures—both at the national and local levels. In Lebanon, the Hizballah has participated in drug dealing that has been at the center of the political chaos over many years. The Tamil Tigers and Kurdish insurrectionists have used drug profits to finance their terrorist activities in Sri Lanka and Turkey, respectively. As examples of repressive regimes that spread terror within their own populations, the war lords of Burma might equally be characterized as drug lords. Pol Pot was closely linked with drugs in Cambodia. In Africa, the Somali war lords are also known to have promoted drug trafficking.[11]

Meanwhile, both the "traditional" and the "high-tech" terrorists in other countries smile over the seeming impotence of the United States to effectively combat drug trafficking. They know and admire the success of the drug cartels in penetrating the borders of the United States, in laundering money, and in establishing strategic networks of criminal elements throughout the United States and its entry points. What better allies could terrorist organizations find than the insiders of the organizational, financial, and contraband infrastructures who have already penetrated America's heartland?

I have repeatedly emphasized the growing links between organized crime and terrorism. Now myriad Russian mafia groups who have added narcotics opportunities to their brand of capitalist ventures dot the international landscape. Thus far, they have not been accused of employing murder and political intimidation directly. They have joined forces with the drug lords of Colombia, and have set up shop in Miami, New York, and San Juan. Russian vessels have off-loaded AK-47 assault rifles and rocket-propelled grenades in Turbo, Colombia, in exchange for drugs. The cocaine they receive is then shipped to European customers. Payments are deposited electronically in any one of the dozen Caribbean banks controlled by Russian criminal groups. Helicopters and surface-to-air missiles are also available for barter, thus providing Colombian narcoguerrillas with new firepower to protect their illicit enterprises. There have also been reports of alliances among Russian criminals and the Sicilian mafia, with the mix of arms and drugs central to their joint activities.[12]

When I recall the international incidents I have witnessed over decades of travel on professional assignments, I can say with confidence that drug trafficking is increasingly spilling into the mainstream of economic life throughout many regions of the world. Although my overseas missions never focus on drug dealers, I repeatedly encounter their telltale signs. Few nations have managed to escape involvement because growers, processors, distributors, and consumers are linked in an ever-growing international money machine.

Americans traveling abroad witness airport guards making their rounds with sniffing dogs at their heels, nervous outbound airline passengers being taken out of line for body searches, distraught inbound arrivals walking by in handcuffs, and rumors in the business lounges concerning apprehension of contraband. Hotel bell captains react in unpredictable ways when I ask for directions to a drugstore. A local's hesitancy when I am inquiring about interesting evening activities can inadvertently lead to an offer of drugs and prostitutes in many European and Asian cities.

In the early 1970s, I traveled to Colombia as an official of the Agency for International Development. I arrived in Medellin to visit Colombian industrial research laboratories serving the region, only to be delayed several hours at the airport. All movement of commercial and private traffic was frozen so as not to interfere with a government raid on the headquarters of the leaders of the Medellin cartel. The Colombian federal forces had arrived by plane from Bogota, and they were paying their call on the reigning drug dealers. Within two hours we saw the federal forces return to the airport, apparently satisfied that all was in order. According to our escort, this was a *pro forma* operation, and the troops never even entered the targeted compound. The kingpins had made the "necessary arrangements" in advance. No knocking down fences. No confrontations. No headlines. The scenario had already become the norm in a country accustomed to bribery and corruption.

On a later trip to Colombia, I flew to Cartegena to investigate the feasibility of conducting ecology studies of industrial developments in the northwest region. While this coastal town continues to be the least drug-infested city in the country, the nearby rural areas, like other regions of Colombia, harbor many drug-related activities. My project depended on Exxon's plan to invest heavily in the coal resources of the region. As part of its arrangement with the Colombian government, the company had agreed to provide large sums of money to local universities to study the surrounding ecology in cooperation with American specialists.

The real interest of local officials, however, had more to do with retaining the thriving tourist trade in Cartegena. They hoped coal production would offer an alternative to the coca business that was rapidly gaining ground in the region and frightening away tourists. If Exxon offered major employment opportunities, then the Colombian farmers might be persuaded to forget about growing or processing coca leaves. Unfortunately, the international coal market simply did not take off as Exxon hoped. Thus, the plan to transform the economic mix of the area also fell flat, leaving the farmers once again at the mercy of their one lucrative crop—coca.

In 1984, I visited Malaysia, a country at the other end of the drug-control spectrum. Ridding the country of heroin en route from the Golden Triangle (Burma, Thailand, Laos) to the markets of the world had become a political passion of the country's rulers. Conviction for possession of fifteen grams or more of heroin meant the mandatory death penalty; only in rare circumstances was amnesty ever to be granted.

As a member of a World Bank team dispatched to develop recommendations for upgrading research and education capabilities of Malaysia, I quickly noted the government's standard welcome to the country. Large billboards on the main streets warned visitors—in Malay, English, and Chinese—that there was zero tolerance for possession of heroin. These billboards repeatedly underscored that in this country "just say no" was not simply a slogan but the most strictly enforced law of the land.

The Hewlett-Packard company had a special interest in the Malaysian market. They supplied the forensic chemistry laboratory of the Malaysian Customs Service with high-precision scientific instrumentation for measuring the exact amount of heroin obtained during drug busts. The key question was whether the quantity exceeded 14.99 grams. If it did, whoever had possessed the heroin would soon be on death row. The laboratory chemists told me that they routinely testified at such trials.

At the time of my visit, I headed the national laboratory of the EPA in Las Vegas. We were responsible for ensuring accurate measurements of chemicals found in hazardous wastes in the United States. We had purchased the most sophisticated instruments available in the world and had received countless commendations for elevating quality control procedures within EPA to the highest levels, namely those that would withstand challenges in American courtrooms.

The Malaysian chemists I met were as competent as my American team. Countless times, they had appeared in court-rooms where their testimonies had been challenged by some of the most powerful figures in Asia. Yet they had developed a set of objective measures of the quality of their work which had re-buffed even the most aggressive and best qualified lawyers.

The chemists reported, however, that work had been taper-ing off. Word was out that the government meant business. Heroin smuggling was no longer a popular occupation. Although there were still some incidents of smuggling along the Malaysian borders, criminals exercised the greatest care, since they were acutely aware of the precision of the scientists who awaited their contraband in the capital city. And these scientists were well paid and had proven themselves to be beyond corruption.

It is, of course, tempting to advocate replication of the Malay-sian model in other countries. However, few developing countries have such reserves of palm oil, timber, and rubber to provide al-ternative economic opportunities for farmers tempted to grow poppies.

During a trip to Vietnam in 1991, I unexpectedly learned how local officials tried to transform addicts who had come under the influence of criminals, into responsible citizens. The initial focal points of the visit of our four-person team were the dramatic in-creases in rice yields in the countryside and the unusual Viet-namese successes in treating, with the support of very modestly equipped hospitals, badly maimed victims of the war with the United States. But the orientation of our visit changed.

In search of agricultural research achievements, we traveled to the Mekong Delta, home of Can Tho University. For thirty years, the school had retained its ties with Michigan State Univer-sity despite the communist subjugation of the country. However, the research program at the university had fallen on hard times. When asked to take our group to a research facility, the local hosts decided to show us a heroin addiction recovery center instead. They wanted to demonstrate to the medical doctor on our team a new technique in the struggle against drug addiction.

Three employees of a small company stirred boiling mixtures of juices from local crops in open vats. The "inventor" of the process explained that his family had practiced herbal medicine for decades. His assistants took the concoction to a nearby clinic where several dozen addicts lay on cots, writhing in agony. They were given doses every few hours. The local doctors claimed that within several days the habit would be miraculously broken, and the patients would be on their way to recovery. Our American doctor, acting through an interpreter, interviewed a number of the addicts who confirmed these lofty claims.

According to the clinic director, the key was then to provide the former addicts with jobs to keep them busy—to keep them away from the criminals with whom they had been hanging out and to focus them on something other than drugs. In the absence of jobs, they would probably return to the habit within a few weeks, he added. We met several graduates of the clinic who were happily employed at a nearby sawmill. They claimed they were drug-free, had never felt better, and were entering a new phase of life.

All the Vietnamese officials testified to the effectiveness of the concoction, though they freely admitted that any side effects, such as cancer or heart disease, had not been studied and were of little concern as long as the potion was effective in breaking the habit. When combined with supervised employment, treatment had proven itself a resounding success. The emphasis on employment opportunities for recovering addicts is certainly a lesson to be applied to our own drug problem. Also, herbal concoctions seem to deserve greater attention, particularly in an age when alternative medicine is viewed more favorably.

Upon our departure, the local officials lamented that they really couldn't do anything to stop the flow of heroin from Cambodia and Laos through their province. Their strategy was to reduce demand and then the contraband would simply transit the region en route to markets elsewhere.

A more recent international initiative revealed a depressing reminder of the lengths to which drug traffickers go in mining a potential new market. In the summer of 1996, I ventured to Bosnia with the objective of developing programs for bringing together the higher educational institutions of the country, even though each was dominated by a different ethnic group. Of all the institutions in the country, the universities seemed best positioned to begin the process of ethnic reconciliation by opening their doors to students of different groups and by cooperating among themselves. E-mail links were an obvious starter, but I also harbored more ambitious plans.

The university in Mostar was a special challenge, for the town was divided. One part of the university was located in the decimated Bosnjak (Muslim) sector and the other in the prospering Croatian sector. I was struck by this vast disparity in economic life and sought an explanation.

According to local officials in the Croatian sector, one word explained much of the difference—drugs. The Croatian sector had become a center for drug traders—savvy businessmen who spent money in the restaurants and hotels, who rebuilt houses, and who employed many farmers and service personnel in their nefarious, but lucrative, trade. Poppies grew nearby, and somehow heroin from Asia also found its way into this war-torn segment of Europe. Heroin, hashish, and pot were in demand by soldiers and to a greater extent by former soldiers who had joined the ranks of the unemployed. Heroin was also in demand in western Europe, and Mostar was a convenient shipping point.

Some Serbian soldiers were condemned by the local officials as having revived Belgrade's tradition, dating back many centuries, of drug running from Turkey. Meanwhile, in Sarajevo a foreign journalist told me that the Bosnian army was bringing drugs into the region on aircraft operated by the United Nations. Whatever the source and whatever the market, there was little hope that the Croatian sector of Mostar would become interested in linking up with the Bosnjak sector—for education or any other purpose. These Croats were happy with their good life and had no intention of experimenting with different concepts of democracy and equality that would threaten their lifestyle.

Finally, in Prague—early 1998, I encountered yet another dimension of narcotrafficking. Czech chemists were being hired by party-givers in London to fly on weekends to England. There, they would synthesize exotic stimulants in small-scale laboratories. Catering for parties had never before been so lucrative.

Yes, narcotics have become a dominant factor of many aspects of international life. International transportation routes are laden with drugs. In certain areas of the world, economic prosperity depends on drugs. To some local officials drugs are a menace, while to others they are a boon to the economies. This indeed is the Catch-22 of drug control efforts the world over.

The drug smuggling infrastructures provide many opportunities for joint ventures in America and elsewhere between traffickers and terrorists with grievances (or simple business interests) —who find the networks useful. A prospering drug trade will almost surely lead to a widening of the international routes used by cocaine-packing criminals of the past into multiple-lane superhighways shared by the armed terrorists of the future.

Returning to the more direct linkages of terrorism and drugs, one of the most obvious examples is the tight web woven between the two in Colombia. With access to huge sums of money from the drug cartels, guerrillas are armed with the best weapons drug money can buy. They have terrorized large segments of the population while ensuring safe passage for traffickers. The relationship provides the smugglers with unimpeded access to rural drug laboratories, airstrips, and coca fields. It allows them to outsource contract murders and kidnappings and gives them access not only to large supplies of weapons but also to new sources of chemicals to process drugs for distribution throughout the Caribbean. Meanwhile, the well-financed guerrillas are able to recruit new mercenaries. They increase the bribes and favors that persuade politicians at all levels to do their bidding, and they expand the territory under their direct control.

The narcoguerrilla agenda is not solely aimed at controlling political and economic life in Colombia, though it is well on its way to meeting this objective. Many leaders also are intent on "getting even with the gringos." Indeed, they become enraged when they hear of comrades extradited from Colombia to the United States. In the 1980s such enmity triggered a particularly bloody massacre at the Supreme Court in Bogota where an extradition treaty was being considered. A number of the nation's leading judges were killed. During 1985 to 1990, the drug cartels financed the killing of one thousand police officers, seventy journalists, sixty judges, and four presidential candidates, with the extradition issue a constant background theme.[13]

The drug dealers derive some measure of revenge for the incarceration of Colombians in American jails by exporting their lethal cocaine and heroin weapons to cities across the United States. Groups with traditional Marxist or Castro-like views of the evils of capitalism (terrorists of the Revolutionary Armed Forces of Colombia or FARC and the National Liberation Army or ELN) target multinational companies that are investing in the country. In 1996, for example, Colombian terrorists carried out more than seventy attacks on oil pipelines and related installations jointly owned by Western companies and the government of Colombia— incidents they viewed as attacks on U.S. targets.[14] And the Colombian president readily acknowledges that terrorist kidnappings, which happen daily, often serve as ways to get even with the Americans.[15] Overall, the Colombian brand of international terrorism is far reaching and very deadly on many fronts.[16]

Closer to home, another type of drug terrorism occurs along the Texas borders. American ranchers, in their pleas for intervention by the U.S. authorities, report that armed Mexican–American drug gangs ride roughshod across property, terrorize families, and disrupt their lives. The gangs have torn down fences on both sides of the border, scattered cattle, commandeered houses, and threatened citizens who get in their way.[17]

The story of Abrego's capture begins to illuminate the scope of the problem as exemplified by the span of the widest cocaine bridge created from Latin America into the United States to evolve during the past decade. More than 70 percent of the cocaine smuggled into the United States in 1996 came through Mexico. (Later we will discuss the gradual shift of the flow to new routes over the coastlines of Puerto Rico.) With more than thirty Colombian and Mexican criminal enterprises in thirty-six American states, Mexican drug cartels have expanded from simple cross-border smugglers to transporters of drugs into a vast network of wholesalers and retailers. According to one estimate, they employ more than one million people; six hundred thousand are in the United States. These numbers are difficult to comprehend but even if somewhat inflated, the enormity of the network is clear—with a growing likelihood of drug dealers and terrorists crossing paths.[18]

Supporting their base of operations in Mexico, drug gangs are known to buy off all the officials they can, and, at times, kill those they can't. While drug bribes are estimated to have skyrocketed to

$500 million per year in Mexico, narcoguerrillas often prefer to eliminate those who oppose corruption.[19] All the while, American law enforcement officials bristle at reports that some maximum security cells of Mexican prisons have been converted into comfortable living quarters, serviced by cooks and maids and equipped with cell phones and saunas. In such plush hideaways, well-connected drug traffickers continue to conduct business unencumbered by inquisitive law enforcement officials.[20]

It is not just our southern neighbors who are driving the drug networks. Opium poppies can grow in almost any temperate country, and 1996 registered a bumper crop in worldwide production. More than half of the crop is in Burma, which single-handedly could meet the world's entire demand for heroin, opium's derivative. The routes from Burma and other South Asian countries to international heroin markets are multifaceted. The economies of Asia, whether up or down, attract ever greater quantities of drugs. China is even experiencing a serious rise in teenage addiction.[21]

The U.S. government takes considerable credit for the demise of Burmese drug lord Khun Sa. In 1995, Sa surrendered his Shan United Army to Burmese authorities, which cut back its activities after considerable pressure from Thailand. While he may have been defanged, his venomous legacy lives on in ways U.S. authorities have long feared, namely, an endless supply route of opium from the Golden Triangle to New York City (with its limitless appetite for heroin) and from there on to other American cities. In the 1990s, Burma's share of the U.S. heroin supply rose to 80 percent.[22]

For many years, Thailand has served as the major conduit for Burmese heroin. Hong Kong, where the massive flow of maritime cargo overwhelms any attempt at meaningful inspection, has been the most popular transshipment point. Large amounts are also traveling overland through China, then out through other Chinese ports. Looking ahead, the political future of Hong Kong is uncertain but its history as a trafficking crossroads will surely live on. It is likely that South Korea, Vietnam, and Taiwan will also be increasingly active as staging points for contraband exports headed to the United States.[23]

Opium production has seized on the imagination of the entrepreneurs in Afghanistan and Pakistan. The wake of the Afghan

War eliminates many other economic alternatives. As for Pakistan, the country has struggled for decades to find ways to ensure economic survival. The insurgent forces within these countries have found that drug profits come in handy and in return have helped ensure that cultivation efforts are not interrupted.

Meanwhile, the disintegration of the Soviet Union has spawned new highways from the opium fields of Asia to the eager consumers of Europe. The opium route has displaced the silk route from China and Afghanistan across the borders into the newly independent Central Asian republics. The new countries with shattered economies, rampant crime, and poorly paid customs and enforcement services offer smuggler-friendly passage. As mentioned earlier, these ancient rails, newly greased by the slick currency of drug dealers and gun runners, are likely *routes for leakages in the opposite direction of nuclear material* or high-tech military weapons out of the former Soviet Union into the volatile political scenes of both Asia and the Middle East. Further to the west, the Caucasus region of the former Soviet Union also provides conduits for the drug trade. Local officials in this region have repeatedly told me of their concern that drug routes might be used for smuggling nuclear materials out of Russia and into irresponsible hands in the Middle East.[24]

The *major* countries of origin of heroin and cocaine are relatively few in number, but many more countries are serving as distribution centers, including Turkey (where Istanbul remains a critical transshipment point), Nigeria, Kenya, Liberia, Angola, and Namibia, to name a few.[25] Traffickers are preying on the still fragile, new democracy of South Africa where, according to one report:

> By land, by sea, by air, cocaine arrives daily, hidden in cargo containers, in compartments of suitcases, in aerosol containers, and in soles of shoes. South Africa has become one of the hottest new transshipment points and domestic markets in the world.[26]

If drugs are so easy to smuggle through these areas, what other types of dangerous contraband could be sent to terrorists?

A particularly troubling development has emerged along our northwest border; drug dealers cluster along Hastings Street in Vancouver trading cash for plastic packets of heroin direct from

Hong Kong, Taiwan, and China. Canada's Pacific paradise has become the drug capital of the country, as Canadian officials boast about the trade boom engulfing the city, including the dramatic increase in truck traffic between the two countries.[27]

Also in Vancouver, Mark Emory, Canada's "Prince of Pot," sells $3 million worth of marijuana seeds annually that then sprout in hundreds of hydroponic indoor gardens in the Province. The Web site for his virtual store "HEMPBC" attracts one million electronic hits per month. The police occasionally undertake raids in the gardens, but usually they defer, claiming other pressing priorities. Emory continues to dutifully pay the $25 tax on each $250 sale of twenty seeds.[28] While a hemp dealer may be a very distant cousin of terrorists, Canadian permissiveness of law breakers at even this level further entices Asian criminal organizations to establish outposts in the front yard of our most important neighbor.

Another type of challenge for America is posed by synthetic drugs, which are staging a major comeback. LSD and amphetamines (speed) are being rediscovered by American youth. The ominous methamphetamine (ecstasy) has a large cult following as well and is rapidly becoming the dominant illegal synthetic drug in the country.[29] The synthetic drug industry is located almost entirely on our home territory. However, there are signs that Mexican sources, with easy access to many of the necessary chemicals, are beginning to enter the methamphetamine trade using their distribution networks throughout the southwestern United States.[30]

Thousands of laboratories pumping out illegal synthetics have been uncovered in California alone during the past decade. Unfortunately, locating these small laboratories is just the beginning of the challenge. The dangers associated with producing precursor chemicals used in synthetic drugs can be severe. Chemicals can explode, burst into flame, or emit caustic fumes. Thus, police and other investigators must use extreme caution in confronting illicit drug production.

As to links to terrorism, the DEA has developed a regulatory and enforcement approach for controlling chemicals illegally used for narcotics. This regulatory experience, albeit not totally successful, can clarify problems and approaches in controlling chemical explosives and perhaps other chemical agents of interest to terrorists. On the more ominous side, the clandestine laboratories that fuel the synthetic drug trade represent a disturbing parallel to laboratories that could be used to manufacture poisonous chemicals, such as the garage-style sarin facility that camouflaged the work of the Aum Shinrikyo cult. Likewise, the capabilities of underground chemists in the United States who have honed their skills in the synthetics drug industry should not escape notice. Granted that the chemical cocktails used to poison, asphyxiate, and otherwise injure victims may be different from those that are used simply to induce psychological highs, the versatility of the basic components of chemistry laboratories raises a worrisome specter indeed.

One measure of the success of the international drug trade is that for the last twenty-five years, each American president has waged a war on drugs. So far, not one of them has come close to proclaiming victory. A succession of American drug czars has been welcomed into the inner circles of the White House amid great expectations. Drug control budgets for more than fifty government agencies have steadily grown. In 1997 these budgets reached $16 billion, a five-fold increase in twelve years.[31] Because of new resources, efforts south of the border by aircraft and patrol ships have expanded. Moreover, procedures at the 300 ports of entry into the United States have been tightened. Foreign government leaders who have profited from the drug trade have been arrested and convicted in the United States, and a few Latin American drug lords have been dethroned and are residing in American jails. Yet the dealers are undaunted.

The importance of reducing the demand for drugs within the United States is often cited as an essential component of any drug control policy, although even a dramatically reduced U.S. demand would not eliminate the criminal infrastructure of the trade and profits from the rest of the world. Meanwhile, the world's appetite for drugs continues to fuel the affinity between terrorists and drug cartels.

It appears that the Clinton administration is devoting greater attention to reducing demand than did the Bush administration, although the 1997 Clinton budget continues to focus two-thirds of the resources on reducing supply.[32] A frequently cited 1994 study by the Rand Corporation concluded that $34 million invested in treatment reduces cocaine use as much as an expenditure of $783 million for programs to destroy crops in Latin America or $366 million for interdiction.[33] Even so, the arguments over whether to address drug supply or drug demand are fruitless and counter-productive. Both must be attacked. Interdicting the supply of drugs is most applicable to attacking the roots of several impor-tant forms of international terrorism, and I have emphasized the supply side of the equation.[34]

That said, America's hard drug of choice remains cocaine for the time being, although heroin is not too far behind; and the ready availability of both drugs shows no sign of decreasing. The selling, buying, and use of crack (the smokable and most highly addictive form of cocaine) fuel much of our inner-city violence and generate most of the drug revenue. While overall cocaine use has dropped markedly since the high consumption of the mid-1980s, it is on the rise again among high school students.[35] Heroin, once considered the drug for down-and-out derelicts, may also be acquiring a dangerous new respectability among young drug users, particularly in the United States. Frequent revelations of use by rock stars and entertainers only add to its caché. Heroin has a special property that appeals to both the addicts and the traffickers. Users often develop a long-term tolerance to certain levels, such that heroin use may be masked beneath the facades of normal lives. For the drug dealers, this translates into a loyal and long-term customer base.

As we have already seen, the narcotics enemy is formidable indeed. The drug operators are as resourceful, sophisticated, effi-cient, and as readily replaceable as the leaders of any multinational corporation. Regardless of the efforts to put them out of business and the minor successes along this road, the steady flow of drugs into this country has continued without so much as a hiccup.[36]

To reduce not only the devastating effects of the drugs them-selves, but the funds available to terrorist groups, we must get a firmer grip on the root of these problems. In principle, reducing the source of cocaine should be relatively easy since it is techni-cally possible to eradicate most of the world's coca crop in just a few months with a relatively benign herbicide. Why isn't this hap-pening? A primary issue continues to be the lack of enthusiasm among the governments of Colombia, Peru, and Bolivia to go head-to-head against both criminal elements and the large population dependent on cocaine income. Why should they when the lives of government ministers and judges are at stake? Why should they when many government officials at all levels are on the payrolls of the cartels?

For more than fifteen years Colombia has reported the highest murder rate in the world. In Mexico, the salaries of those guarding cocaine operations are ten times the salaries of the police and soldiers who should be on their cases. Whether driven by fear or conviction, or both, many politicians in Latin America consider the growing of coca to be a birthright—the birthright of the peasant who has no other means of subsistence. This theme is not lost at election time.

Despite periodic eradication campaigns, the cocaine trade has continued to grow. While exports from some countries may decline and production in certain regions may be terminated, overall trends are not in decline. Even the heavy investment by the United States and others in promoting alternative crops, particularly pineapples and bananas, has had limited impact in reducing overall supply.

Of course, firmly entrenched between the Latin American farmer and the street retailer are the refiners, the shippers, and the wholesalers. The only effective strategy is a comprehensive blitzkrieg of the entire chain to cauterize the dealers' networks wherever they are exposed and to choke off the money supply to their symbiotic terrorist groups. Unfortunately, the infrastructures serving the drug trade exhibit few vulnerabilities. Indeed, the mastery of the complexities by the heads of the operations can only be described as remarkable.

To appreciate what the traffickers have accomplished, consider the intricacies of moving everyday food products from a farm in Iowa to the supermarkets of the United States, and the world. Careful planting and cultivation, controlled harvesting and storage, well-organized transportation and processing, and dependable delivery in response to customer schedules are the essential steps. Similarly, financial arrangements must be carefully laid out from financing each step, to agreeing on prices and means of payment, to transferring the funds to the many involved organizations through a variety of mechanisms. In the case of food products, all activities are undertaken within a completely legal environment, with strong support along the way from both state and federal agriculture agencies.[37]

Likewise, the drug traffickers must oversee each step of the production/delivery and the financing/payment chains. But they must operate on a tightrope swinging outside the edges of the law. Governments foresworn to prevent their operations, police forces and armies searching for their hideouts, and international law enforcement agencies ready to pounce on them weave a series of nets under their high-wire act, hoping to benefit from the most minor misstep. Like food growers, the narcotics bosses are expected to deliver quality goods. Therefore, some suppliers provide money-back warranties on the purity of their products while others offer free samples to new customers.[38]

One test of the effectiveness of U.S. programs is what comes across the border. The picture is not pretty. Senator John Kerry described what is happening in cocaine imports:

> Cocaine is hidden in the walls and support beams of cargo containers and within bulk shipments of coffee, frozen inside vast quantities of blast-frozen shrimp to deter drug-sniffing dogs, and even moved in submarine-like submersible ships to avoid detection by conventional radar and sonar. Cocaine arrives concealed in beach towels, inside spools of industrial thread, inside cans of lard, sealed with quartz crystals and in drums of fruit pulp, in avocado paste, in fish, and in condoms stuffed into the intestines of boa constrictors. One trafficker poured liquid cocaine into a shipment of live tropical fish, keeping the fish protected in an inner bag of water.[39]

In 1997, the Clinton Administration heralded its successes in combating this awesome challenge, in weakening the drug trafficking organizations at all levels, and in seizing their assets and products as follows:

> We made solid gains against the drug trade in 1996. Working with our allies in the Western Hemisphere, the U.S. government-led efforts pressed the drug syndicates at their most vulnerable points. We interrupted their preferred trafficking approaches to the United States, forcing them to shift to longer routes in the eastern Caribbean. We helped the drug source countries to eradicate coca plantations and opium poppy fields. The year's most successful operation was the disruption of the

air-bridge that carried the bulk of Peruvian cocaine base to Colombia for processing and distribution. The cut-off not only deprived the Colombian trade of essential basic materials, but so depressed the price of coca leaf in Peru that growers abandoned fields in the coca-rich Upper Huallaga Valley. The exodus lowered Peru's coca cultivation in 1996 by 18 percent to the lowest levels in a decade. This was a significant change for the world's largest coca-growing country.[40]

Rounding out the positive aspects of slowing the flow of drugs, the Clinton administration added the following:

> Small steps do not grab headlines; many never come to the attention of the media. The countless routine drug seizures, the jungle drug labs or airstrips destroyed every day, the arrest of corrupt officials, or the improved performance of police and judicial authorities benefitting from U.S. government assistance receive at best only fragmentary coverage in the world media. But these same small steps add up to important and lasting gains at the expense of the drug trade. As we have seen, cumulative achievements pay off. Over the long term, such steady progress offers the best hope for transforming a potential threat to the stability of nations into a manageable nuisance.[41]

All of us look forward to the days when drug trafficking can be classified as a "nuisance" and no longer a menace. In the meantime, our achievements may be laudable, but we continue to nibble at the edges of a huge problem. The head of the U.S. Customs Service has rightly asked, "While it is splendid that we intercepted a million pounds of narcotics in 1996, how can we stop the flow in the first place?"[42]

Indeed, how can the American public have confidence that we are even on the right course? Despite all the positive reports of progress, we find President Clinton's drug czar turning over confidential information to his Mexican counterpart whom we later discover is on the payroll of the drug cartels.[43] The American general responsible for South American operations pleads for funds to sustain an army in Colombia that has little will to fight.[44] American helicopters are provided to spray marijuana fields with herbicides but end up scattering fertilizer instead.[45] The U.S. government refuses to allow a single Colombian drug to lord go free in exchange for purportedly reliable information on the details of the smuggling routes from Colombia to the United States. Along these routes flow 80 percent of the contraband cocaine heading north. The information would implicate the Colombian president as a party to the illicit trade, according to public reports.[46] Most astonishing, the director of the CIA must go to the Watts region in Los Angeles to convince a hostile gathering of residents that his agency is not subsidizing the drug flow from Central America into Los Angeles.[47]

Perhaps the most worrisome development of all is the increasing numbers of U.S. customs, drug enforcement, border control, and immigration officials who have been sucked into the world of corruption along our borders. The temptation to double an annual salary in return for an hour's inattention to duty is understandably very great indeed.[48] A recent report by the inspector general of the Department of Justice highlights a few examples of U.S. government employees lining their pockets from drug corruption on the Mexican border. Greed to support a lavish lifestyle or to offset dire financial need has resulted in Border Patrol agents waving drugs through checkpoints, immigration inspectors providing temporary Resident Alien cards to drug runners, and other inspectors actually directing contraband traffic across the border. These cases, though fortunately still very small in number, are hardly confidence builders.[49]

On the other side of the border, the report of an entire Mexican special drug-interception unit being arrested for using its plane to smuggle cocaine in from Guatemala just adds to the mistrust of law enforcement officers. Trained by the U.S. Customs Service, these renegades were caught by pure chance when drug-sniffing dogs, taken aboard the aircraft for a demonstration, detected the contraband and went into a frenzy.[50]

Meanwhile, the American press reports continuing turf wars among agencies in the corridors of Washington. For example, eleven agencies have established counternarcotics intelligence centers around the country, with some of the databases off limits to other agencies. One joke circulating around Washington goes:

> What happens when three police dogs are in a room containing a gun and some hidden drugs?
>
> The dog trained by the DEA finds the drugs, the dog trained by the ATF finds the gun, and the dog trained by the FBI

holds a press conference to announce that the FBI has seized a gun and a kilogram of drugs.[51]

We welcome the news that scientists have developed x-ray, gamma ray, and positron emission technology that can tell whether a truck is carrying narcotics, a nuclear weapon, or a head of broccoli.[52] It is essential that such technology not only be deployed as soon as possible but be in the hands of officials who, regardless of their organizational affiliations, can seize the transporters whether they work for a drug cartel, a terrorist organization, or an importer whose agricultural products have become infested with insects.

In looking to the future, we should build on the extensive efforts, and indeed the many successes cited below to disrupt drug trafficking. The problem may seem overwhelming, and there is little time to reinvent tried and true approaches even though past efforts have clearly been inadequate.

International agreements: American diplomats have for years played leadership roles at many international forums in developing treaties and agreements that strengthen efforts in countering the drug trade while also imparting ever-increasing consistency into the regulatory and enforcement efforts of different nations. Conventions to fight narcotics and money laundering have been pushed within the framework of the United Nations. Agreements on anticorruption and drug trafficking have been reached at summit meetings with presidents from North and South America. A host of other international accords have been put into effect. Of course, the reach of these international obligations needs to encompass more countries. Vague provisions should be nailed down. Implementation of commitments must replace words with deeds. While paper agreements will not stop narcotrade, they provide even in their current form an important international standard for judging the behavior of all countries. They are the starting point for forceful actions against those who disregard the will of the international community.[53] They provide impetus for international sharing of intelligence information through Interpol and other organizations.

Interdiction: At the operational levels of interdicting the flow of drugs headed for the U.S. southern border, the United States has entered into bilateral agreements with many Latin American countries. They provide, for example, for American authorities to board suspicious ships, to pursue drug runners across national boundaries, to overfly the airspace of cooperating countries, and to order planes bearing contraband to land without regard to violating airspace.[54]

Turning to the immediate problems at U.S. ports of entry, the U.S. Customs Service, the Border Patrol, and other agencies have developed many measures for uncovering hidden contraband. The volume of incoming goods—by ship, air, and land—is staggering. Only a small fraction of the potential carriers of hidden cargo can be checked. With 6,500 inspectors, the Customs Service must rely increasingly on technology to ease the task. Portable range finders locate false walls and concealed compartments in storage containers. Fiber optic scopes permit visual examination of gas tanks and enclosed spaces. Inspectors wield hammers, drills, and pry bars to open containers designed to thwart inspections. They use steel probes to detect drugs hidden in flower baskets and in containers full of bulk materials. They deploy 78 small vessels and 115 airplanes and helicopters to help in their searches for those who skirt customs check points.[55]

The heroics of the Coast Guard in intercepting maritime contraband are frequently reported on television. The sight of a Coast Guard cutter signals bad news for a smuggler. The Coast Guard's cooperation with the Mexican navy is growing stronger every year, and its new focus on preventing the offloading of drugs in Puerto Rico is critical since there are no barriers between San Juan and the U.S. mainland.[56] Indeed, Puerto Rico may soon rival the Mexican border as the most important entry point of drugs coming into the United States. In 1997, 40 percent of the cocaine entering the United States passed through Puerto Rico as compared to 30 percent the year before. With seventy-two flights to the American mainland every day and San Juan the busiest port in the Caribbean, the coastline of Puerto Rico must receive greater attention than the currently expanded Coast Guard efforts can provide.[57]

The Department of Defense is also an important player in interdicting drugs en route to the United States. The department's recent activities have included assisting Peru and Colombia in destroying drug-running aircraft, supporting a new counterdrug command center in Thailand, providing radar equipment for drug interception missions in Central America, and expanding intelligence support for U.S. agencies. Also, special units of the National Guard assist the Customs Service in inspections of cargoes and seizures of drugs. However, any additional role of the Department of Defense is a matter of considerable controversy. Supporters claim it is foolish not to fully utilize the capabilities of the Pentagon, whereas opponents argue that military personnel are not appropriately trained for the task and that such deployment would be a diversion from the Pentagon's core missions.[58] But the muscle of the Department of Defense sends a message to the drug lords that the United States is dead serious in defending its borders from any type of illegal assaults.

Pressing for Reduction in Supply: The experience of U.S. agencies in encouraging and supporting efforts of countries to reduce the supply of coca and heroin, while not very successful, is nevertheless very important. The efforts have been largely of three types:

- Training and equipment for local law enforcement organizations with responsibility for controlling the narcotrade;
- Destruction of crops and efforts to find alternate crops;
- Sanctions and other economic punishment when countries do not exert enough effort to reduce the flow of narcotics.

The scorecard for these efforts is mixed, with successes claimed in Thailand, for example, but primarily frustrations reported in Latin America.

Of special interest has been the *certification* process. Each year on March 1, the U.S. president reports to the Congress on whether those countries known to be illicit drug producing or drug transit countries have taken appropriate steps to attack narcotrafficking. He can certify that a country is cooperating with the United States in this regard or is taking appropriate action on its own to meet the antitrafficking goals specified in the 1988 UN Drug Convention. He can decertify a country as not cooperating and not taking unilateral action or grant an uncooperative country a waiver from retaliatory action if the U.S. national interest so warrants. (See Table 12.)

TABLE 12.   1998 Annual Certification Report

(Major illicit drug producing and transiting countries eligible to receive U.S. foreign aid and other economic/trade benefits)

Certified as fully cooperating (though not all are aid recipients): Aruba, The Bahamas, Belize, Bolivia, Brazil, China, Dominican Republic, Ecuador, Guatemala, Haiti, Hong Kong, India, Jamaica, Laos, Malaysia, Mexico, Panama, Peru, Taiwan, Thailand, Venezuela, and Vietnam.

Not fully cooperating (but eligible for continued U.S. assistance because it was deemed in the U.S. national interest): Cambodia, Colombia, Pakistan, and Paraguay.

Decertified and made ineligible for assistance: Afghanistan, Burma, Iran, and Nigeria.

Source: *International Narcotics Control Strategy Report*, U.S. Department of State, March 1998, pp. xiii–xiv.

In the event of decertification, a cutoff of most bilateral foreign aid is mandatory. The United States must also vote against loans to the country from international development banks such as the World Bank, and ensure that country is denied a quota to export sugar to the United States. In addition, the president may take additional steps such as raising tariffs on imports from that country and curtailing air traffic between the country and the United States.

There is disagreement among experts as to the economic impact of such sanctions as we will see later. However, the political stigma of decertification is serious; it sends a signal around the world about the irresponsible behavior of that country. Most governments would like to avoid such a branding.[59]

Often the debate about certification revolves around Colombia and Mexico. In Colombia, U.S. officials regularly argue that we cannot afford a reduction of any type of aid that prevents honest and cooperative people from combatting the problem, directly or indirectly. The argument is that much of the government may be corrupt, but the business community and the army are supportive of drug control and therefore deserve our support. Opponents point out that the army is unreliable with a poor human rights record and that the corrupt government will not seriously tackle the narcotics problem. In Mexico, the arguments are even more heated. Advocates of certification say that we cannot afford to sour political relations when we have so much at stake in that country. They contend it is unrealistic to expect Mexico to sacrifice an economic activity that returns $30 billion annually to the country. Opponents argue that simply tolerating a poor record makes a sham out of the certification process. They point out that the Mexican government not only consistently works with the major drug-trafficking families but allows paramilitary drug organizations to use foreign mercenaries from Israel, Great Britain, and Colombia as trainers, advisers, and members. In 1997 the president decertified Colombia and certified Mexico. In 1998, the President granted Mexico full certification and granted Colombia certification, not because it met the certification criteria, but because he determined there were overriding national interest reasons to override shortcomings in combatting narcotrafficking.[60]

The vast array of counter-narcotics programs already in place around the globe will continue to have an impact in making narcotrafficking a more difficult and risky occupation. It is clear that a reactive policy based simply on responding to the ingenuity of the drug dealers will bring us no closer to choking off the roots that feed the industry.

For the near term, the political rhetoric of America's leaders is strong. The general strategies for fighting the drug war seem appropriate. It is the forcefulness of the implementation that is the problem—as we hesitate to create diplomatic waves, as we show great reluctance to jeopardize the foreign interests of American business, as we squabble over deployment of American military assets, and as we refuse to complicate border crossings for Americans even though permissiveness benefits illegal traffickers.

The lofty U.S. goals for cutting off the drug supplies (see Table 13) are simply not matched with sufficiently aggressive programs to make a real difference. Only a full-court press on a global scale, involving all of our traditional allies, against the drug lords and their criminal associates will bring us closer to a global political and economic lifestyle worthy of the next century.

Having said all that, there are a few lights at the end of the very long tunnel to the future of drug control. They should be used as guidelamps to expedite our moves forward.

Partnerships: A promising new approach in the United States and other countries calls for private sector importers to join forces with U.S. government agencies in ensuring that goods addressed

TABLE 13.   Strategic Goals and Objectives
1997 National Drug Control Strategy

Goal 1: Educate and enable America's youth to reject illegal drugs as well as alcohol and tobacco . . .

Goal 2: Increase the safety of America's citizens by substantially reducing drug-related crime and violence . . .

Goal 3: Reduce health and social costs to the public of illegal drug use . . .

Goal 4: Shield America's air, land, and sea frontiers from the drug threat.
- Conduct flexible operations to detect, disrupt, deter, and seize illegal drugs in transit to the United States and at U.S. borders.
- Improve the coordination and effectiveness of U.S. drug law enforcement programs, with particular emphasis on the southwest border, Puerto Rico, and the U.S. Virgin Islands.
- Improve bilateral and regional cooperation with Mexico as well as other cocaine and heroine transit zone countries in order to reduce the flow of illegal drugs into the United States.
- Support and highlight research and technology, including the development of scientific information and data, to detect disrupt, deter, and seize illegal drugs in transit to the United States and at U.S. borders.

Goal 5: Break foreign and domestic drug sources of supply.
- Produce a net reduction in the worldwide cultivation of coca, opium, and marijuana and in the production of other illegal drugs, especially methamphetamine.
- Disrupt and dismantle major international drug trafficking organizations and arrest, prosecute, and incarcerate their leaders.
- Support and complement source country drug control efforts and strengthen source country political will and drug control capabilities
- Develop and support bilateral, regional, and multilateral initiatives and mobilize international organizational efforts against all aspects of illegal drug production, trafficking, and abuse.
- Promote international policies and laws that deter money laundering and facilitate anti-money laundering investigations as well as seizure of associated assets.
- Support and highlight research and technology, including the development of scientific data, to reduce the worldwide supply of illegal drugs.

Source: *The National Drug Control Strategy*, 1997, The White House, February 1997.

to them are not concealing drugs. The Mattel Company, for example, has initiated a program of allowing only known, reliable firms to pack its containers in the countries of origin and to load and transport the containers all the way to U.S. shores. The company is determined to keep their toys coming from Asia from being used to camouflage heroin shipments, which are extracted before reaching America's shopping malls. Other companies are also concerned that their goods en route to the United States not be tarnished with the fingerprints of drug dealers, and have joined in similar programs of cooperation with government agencies.[61]

Attacks on Corruption: The governments of the source countries must take hold of the drug problem in all of its dimensions, recognizing that sustainable economic development is the only viable solution. Providing alternatives to the drug trade, while gradually stamping it out, is the ultimate challenge. Under steady pressure from U.S. authorities, the governments of Latin America have finally agreed (in principle) that effective democracy and sustainable development require a comprehensive attack on corruption, with the drug trade being the most pervasive stimulant of corrupt practices. But what other position could they take publicly? We can only hope that the leaders of the key states are truly committed to effective programs. The courage of some government officials in Colombia and Mexico in arresting powerful drug lords at considerable personal risk perhaps represents a gradual shift away from the laissez-faire attitudes of the past.[62]

Unfortunately, to date such heroics have been the exception and represent merely one or two leaves of coca in acres and acres of fields across the planet. But whether it be leaf-by-leaf, acre-by-acre, or country-by-country, we have no choice but to press forward to vigorously combat this blight.

Bigger Sticks and Tastier Carrots: The U.S. government should be much clearer and more credible when it comes to the stick it wields and the carrots it offers. When combined, there is a chance that many drug dealers will be out of business in several decades and that new entrants will be limited in number.

The certification process is a big stick. It is flexible, applied on a country-by-country basis, and can carry considerable weight in terms of political impact. Many other countries, particularly in Europe and Asia, must share U.S. commitments to combatting narco-trafficking: they too can reduce bilateral aid, raise their tariffs, and cast their negative votes in the international development banks to add more economic muscle to the process of decertification.

As to the carrot, new jobs for farmers now dependent on the drug trade, legitimate employment opportunities for entrepreneurs who have used drugs as their vehicle for accumulating

wealth, and better salaries for government officials seem essential
if people are to shift to less threatening occupations. We need to
face up to the challenge very directly: significant numbers of jobs
can be created in the poor countries only by massive infusions of
resources from abroad. If these resources are portrayed as foreign
aid, the proposals will never clear Capitol Hill. However, if devised as leveraged cooperative programs to reduce narcotrafficking, the chances for support are much better.

I will discuss later the details of such expanded programs,
recognizing that collaborative ventures must address not only
drug issues but also the related topics of money laundering, terrorism, organized crime, and corruption. It is clear that jobs must
be a central element of realistic and long-term programs directed
to the root causes of the attraction to and the roaring success of the
worldwide drug trade.

Finally, in the near term, there are unrealized possibilities for
applying strategies learned in battling drugs to combatting other
types of terrorist threats. As we have seen, techniques to uncover
illicit drug laboratories could be useful in finding laboratories for
chemoterrorism. Airport procedures to reveal hidden narcotics
could help discover even more dangerous materials. Postal
screening for drugs has relevance to postal screening for chemical
or biological materials.[63] Just as the terrorists will lean on the drug
dealers for help, the counterterrorists should load some of their
weapons with ammunition developed by those who combat
drugs on a daily basis.

CHAPTER 7

Laundered Assets: Cleaning Dirty Currency to Fuel the Terrorist Engine

The laundering process allows narcotics traffickers, terrorists, perpetrators of financial fraud, and every other criminal enterprise to perpetuate, and to live lavishly from, their illegal activity. Money laundering amounts to hundreds of billions of dollars annually.
Financial Action Task Force
(representatives of 26 nations), 1997

Money launderers hate cash. Shipping huge wads of banknotes is a logistical nightmare. Transferring money electronically is both easier and quicker.
The Economist, July 1997

Offshore banks have become nothing more than partitions on a computer hard drive. Meanwhile, $6 trillion have found their way to offshore havens.
American Expert on Financial Fraud, 1997

*A*n American lawyer listened intently as he sized up his Caribbean host. They sat in an office far removed from the lawyer's more familiar surroundings in Washington, D.C. The chairman of the Bank of Georgetown in the Cayman Islands (BGCI) was explaining to his guest the ease of moving money out of U.S. banks. The American would write checks to corporations BGCI had set up in the United States for receiving money from people who wanted to obscure the money trail. Then, through a Cayman corporation, with a completely concealed ownership, the customer could open brokerage accounts and trade on stock markets around the world. Finally, he could bring money back into the United States with no trace of its source.

The chairman offered a gold Mastercard for his client's private use. The card had no name on it. The banker assured him he could use it anywhere in America, writing in any name he desired, to obtain fast cash or make purchases. Through some slight-of-plastic maneuver there would be no record of the ownership of the card or of the transactions conducted with it.

"And what if I want to bring a large amount of money back into the United States? The gold card won't work for that, will it?" the skeptical lawyer asked. "Don't worry," his monetary mentor said confidently, "We'll move money from your corporate account in the Caymans to a bank in Europe, which will then lend the money back to you. And when you pay interest on that account, you, of course, will deduct it on your U.S. income tax return."[1]

This meeting between an unsuspecting bank chairman and a disguised American consultant to the British Broadcasting Company (BBC) took place in 1994. Similar discussions had occurred thousands of times involving real customers, i.e., criminals rather than journalists. They will continue not only in Georgetown but in rapidly multiplying financial sanctuaries in other countries throughout the world.[2]

Once the hideout for Blackbeard the pirate, the islands of Grand Cayman, Cayman Brac, and Little Cayman have never shed their historical roles of harboring funds of dubious origin. They have been known as "the Geneva of the Caribbean," referring to decades past when they assisted the Swiss in washing money whiter than any other launderers. Now, in the capital city of Georgetown, 550 banks operate with assets in excess of $400 billion. Most banks have been described as brass plaques with no tellers or vaults. Fewer than 15 percent of the banks have ever seen any cash.[3]

What exactly is money laundering? According to the *U.N. Convention Against Illicit Traffic in Narcotic Drugs and Psychotropic Substances of 1988*:

Money laundering involves the conversion of illicit cash to another asset, the concealment of the true source of the illegally acquired proceeds, and the creation of the perceptions of
legitimacy of source and ownership.

In other words, it forever hides the source of income from the prying eyes of the Internal Revenue Service, the Drug Enforcement
Administration, and other regulators. E-cash is now the most common form of currency flowing through the financial laundromats.

How does the money trail lead to terrorism? We have already
seen the interlocks between the drug cartels and terrorists—two
tentacles of a leviathan insatiable in its appetite for money and
prepared to do battle whenever necessary. The drug trade generates more than $200 billion per year, money that is washed
through many different cycles, with a substantial cut ending up in
the hands of terrorists. The symbiotic relationship between drug
running and political violence is nowhere more obvious than in
Colombia, but as already discussed, such ties have also penetrated
other areas around the globe.

Beyond the drug connection, traditional organized crime
groups (the Italian Mafia, the Chinese Triads, and the Japanese
Yakuzas not to mention the newer Russian mafias and Nigerian
bandits) are also finding common interests with terrorist groups.
In some respects these alliances are becoming hybrid melting pot
operations. Whether on the generating end or the receiving end of
funds, and whatever their motivations, constituencies of the hybrids are focused on illicit money; money derived through unlawful means must be disguised and washed into legitimacy.

This is not a new practice. Organized crime groups have long
been at the center of the laundering process. While many wealthy
individuals avoid taxes by sending large sums of money to silent
financial havens, most laundered money is tied in some way to
ventures of drug cartels and other large criminal organizations. Of
course, there is a measure of overlap; some of the wealthy individuals have accumulated their fortunes through connections
with organized crime.

Money is critical for terrorist operations of any scale. In particular, large terrorist organizations operating with minimal or no
state sponsorship usually require substantial sums on a regular ba-

sis. Even small-scale operators must pay for weapons and for airplane and train tickets, as well as maintenance and feeding of their followers. They may resort to counterfeiting and to petty theft, but they still have to move the money rapidly. If creating a major incident in the United States is the terrorist's objective, large quantities of money beyond those concealed in suitcases will likely be required. American law enforcement officials proclaim that they are determined to ensure that there will be no easy entry points in the United States for the money that fuels such evil plots.[4]

As one example, the Aum Shinrikyo is reported to have accumulated prior to the Tokyo attack more than $1 billion, and possibly up to $2 billion when the value of their worldwide real estate is included.[5] Surely, they were not interested in revealing their income sources or the expenditures of these funds. Some of the funds were derived from narcotrafficking. The Provisional Irish Republican Army, the Hizballah, the Tamil Tigers, and other terrorist organizations for years have knocked on many doors in the United States and in Europe searching for funds to be transferred to their accounts, sometimes pleading the need for support for the widows and children of victims of atrocities abroad and sometimes simply concealing their motives. Now with these groups officially designated by the State Department as "terrorists," they probably will have much more difficulty moving money out of the United States.[6] Money managers in India have been hard at work for several decades spinning money through British laundromats. The cleaned currency is used to finance terrorist violence of Sikh and Kashmiri secessionists back home.[7]

Two additional examples reinforce the contention that terrorists are in the money laundering business. Asian banks have erased the money trail of funds used by the NPA terrorist "sparrow squads" of the Philippines, and yet-to-be-identified Caribbean banks cleaned millions of dollars from the Colombia M19 (April 19) movement during the height of its terrorist acts.[8]

As discussed, the drug trade generates the largest amount of contraband cash of any single type of activity—an estimated $200 billion worldwide each year. There is another $400 billion or more being generated through all types of fraudulent dealings with no direct ties to drugs, propeling money laundering into third place among the world's leading industries—after currency exchange and automobile sales.

Of special interest to terrorists in the Middle East have been counterfeit schemes targeted on American currency, particularly $100 bills which have become the common denominator for currency in crime circles. For example, in Lebanon, during the 1980s counterfeiters controlled by Syria and Iran once turned out as much as $1 billion in high-quality $100 bills. These operations led Senator Patrick Leahy, a close watcher of high-tech terrorism, to observe in 1994:

> These are relatively poor countries. Are they going to use this money to buy sophisticated weaponry? Will they try to destabilize our currency? Will they use this to fund smaller scale, but still serious, terrorist activities throughout the world? Think what a terrorist organization, state-sponsored and state-funded with what appears to be U.S. money, could do in purchasing weapons-grade plutonium, especially today, or chemical or biological weaponry.[9]

The illicit counterfeiting campaign in Lebanon was one of the triggers that led to changing the design and composition of $100 bills. There was the hope that if the old bills could be phased out quickly, many launderers would be left holding bags full of tainted bills that would not respond to the usual detergents. Many American groups with extensive activities abroad opposed the proposal to quickly invalidate the old bills. Their argument was based on the degree to which a hasty transition would complicate their legitimate transactions overseas, so this strategy did not become a reality.[10]

Senator Leahy also advocated the use of $100 bills of two different colors: one for use in the United States and one for use abroad. The idea was to complicate the efforts of the launderers who carry U.S. currency abroad and then deposit the contraband money with greater ease in foreign banks than they can in American-based banks. This time, opposition from the Federal Reserve, which had grown comfortable over many years with the use of a single bill, defeated the scheme.[11] Although we will see more successful attempts to inhibit money laundering schemes, U.S. $100 bills will no doubt continue to be subject to manipulation and mimicry.

In sum, for years many loosely associated money handlers around the world have made vast fortunes on boutique fraud schemes, designed to meet specific needs. These operatives, while still in full swing, have been joined by larger groups of mobsters under highly skilled management. They view money laundering not only as the venue for an opulent lifestyle but also as a way to finance more violent crimes. High-speed yachts, luxury cars, and expansive estates are not necessarily the principal rewards for the money gained illicitly. The profits are increasingly reinvested into businesses driven in part by political agendas and armed with violent tactics.[12]

According to President Clinton's drug czar, General Barry McCaffrey, in 1997, each of the over three million deeply addicted drug users in the United States spends an average of $250 per week to feed their habit.[13] In addition, millions of other more casual users are involved in transactions of $10, $20, $50, and larger on a regular basis. As a consequence, the drug traffickers need to take more than $60 billion out of the United States each year.[14]

Cash is not so easy to dispose of, at least in the United States. To move it around the world poses a number of challenges. In $100 bills, the cash weighs three times the cocaine that generates it. In the street-level denomination of $20 bills, it outweighs cocaine by fifteen to one. It is not uncommon for dealers to burn one and five dollar bills rather than be burdened by such bulky currency. In this business, storing small bills in the mattress could be a risky option.[15]

Twenty or thirty years ago, money launderers were simply couriers who picked up the funds from street-corner dealers and deposited them in the nearest banks. As the U.S. government tightened regulations concerning the reporting of bank transactions of $10 thousand or more, the proceeds were packaged in bundles of $9 thousand. In one case, $29 million was transferred to Ecuador in four thousand installments.[16] Finally, the U.S. government cracked down on bank lots of less than $10 thousand as well, and the money manipulators were suddenly faced with the necessity of becoming more sophisticated in their transactions.

An easy transition was to pass the money through institutions legitimately conducting other types of international business: travel agencies, currency exchange dealers, and international brokers. Another technique was to work through car dealerships, brokers of insurance annuities, and check-cashing stores. As the law enforcement agencies systematically choked off these conduits, still additional approaches were put in place.

A highly publicized case in the mid-1990s involved twelve New York businesses, employing 1,600 agents, that wired $1.2 billion per year to South America. These transfers included hundreds of millions of dollars of illegal drug proceeds headed for Colombian cartels. While the companies were not publicly identified, they were described as having businesses ranging from pager sales outlets to travel agencies to mom-and-pop convenience stores. Wire-transfer companies provide a valuable service in low-income communities which may not have easy access to banks, but as demonstrated they also have served home-grown laundering operations.[17]

In the federal crackdown in New York, the transaction threshold required for reporting wire transfers was lowered from $10 thousand to $750, resulting in an immediate drop in funds sent to Colombia. Quickly, smugglers tried to pick up the slack and became burdened down with $100 bills en route to Colombia; cash seizures along the eastern seaboard began to increase markedly, up from $7 million in a three-month period in 1995 to $29 million after the crackdown.[18]

The U.S. government has erected many barricades to hold back the financial floods headed south and in other directions. Clamping down on American banks and bringing other financial institutions into the regulatory net have resulted in a significant increase in laundry charges to criminals. In 1988, when money laundering became a federal crime, washing money cost an average of about 6 percent of the total value laundered. The price had increased to as much as 26 percent by 1996. The process has become sufficiently complicated that the drug cartels usually outsource the laundering of their proceeds in the same way they outsource the drug transportation. The money launderers in turn operate like stock brokers looking for the best deals. These brokers simply offer dirty money to the highest bidder. Since the brokers usually commit to a fixed price with the cartels, their own profits depend on the deals they arrange.[19]

In 1992 the Drug Enforcement Administration learned of a five-stage process used at that time by money launderers in Europe and probably reflective of many current schemes. (See Table 14.) They collared a Colombian in Luxembourg whose records revealed the intricate networks between 115 bank accounts in sixteen countries, creating a successful infrastructure for international money laundering, as shown below. The initial deposit is the key. As enforcement of U.S. laws tightens, the difficulty of using the U.S. banking system at this early stage increases. Assuming a successful deposit, the cash must leave the United States or other countries where laws are generally enforced, but for what destination? As we have seen, the banking megaplex in the Cayman Islands leads a long list of eager recipients.[20]

Another growing concern relates to the globalization of the security markets. The United States attracts much more capital into its markets than flows out, since American markets are considered the most liquid, the most efficient, and the safest. While it is certainly desirable for capital to move across borders freely, steps by law enforcement authorities are necessary to curtail the sanitizing of proceeds from crimes by circulating the proceeds through our markets.

Also, beneficiaries of drug or other illicit money may be buying their way with laundered money into legitimate American

TABLE 14.   A Well-Honed Approach to Money Laundering

- The first step is the initial deposit, which must be made to a bank in a country where the launderer knows he and his associates will not be arrested within twenty-four hours and the money cannot be frozen quickly. The deposit is the single most important step where the money is the dirtiest, where it is most directly tied to the illegal source and therefore subject to seizure or forfeiture.
- The succeeding stages are complex, but increasingly mechanical. In the second stage, the money is transferred to a bank by a non-Latin, usually Spanish, company. Next it is transferred to an account in the name of a Japanese or West European company. Then, once processed there, it can be put either in a working account, most frequently in Colombia or in a savings or investment account in Europe or the United States. In Colombia, the final stage is conversion into Colombian pesos.
- This series of transactions serves three purposes: it creates a complex paper trail, makes the origin and ownership of money dubious; and comingles drug money with legitimate financial transactions.

Source: David A. Andelman, The Drug Cartel's Weak Link, *Foreign Affairs*, July/August 1994, p. 101.

businesses. They can invest and become owners of debt and of re-payable bonds in American capital markets, for example. This can be done through multinational security firms that follow U.S. reg-ulations to the letter in the United States—but whose overseas of-fices accept what have been described as "bags of money" in Latin America, Asia, or Africa where such regulations do not apply.[21]

Finally, criminals use dirty money to purchase equity posi-tions in international banks. Then, as partners guiding the opera-tions of the banks, they are in good positions to encourage bank policies and procedures that work to their advantage.[22]

Evil financial barons, with little regard for the lives or assets of others, may have major outlets in the Caymans and other dis-tant enclaves. But they and their collaborators also frequently in-filtrate governments, occupy board rooms, and prowl the streets in cities throughout the world. They influence decisions that affect the well-being and security of all of us, and it is in our best inter-ests to continue national and international efforts to close in on their illicit empires.

During interviews conducted with two hundred computer sleuths who had targeted businesses and financial institutions, re-searchers uncovered a new dimension of the problem of protect-ing financial assets. The criminals confessed they often faced the dilemma of determining how much money they should steal elec-tronically. According to this research, embezzlers in large compa-nies worry that if they steal too much money, they will surely be caught. If they do not steal enough, the lack of sufficient funds to travel freely and hire good accountants and lawyers will also re-sult in their capture. Logically then, criminals should feel com-fortable in making illicit electronic transfers on the order of $23 million per transaction, the average size of a legitimate fund trans-fer by a large company.[23]

The technical problems of intercepting illegal electronic money transfers by thieves or laundrymen are enormous, and the volume of such e-cash out of the United States dwarfs the amount of money carried across the border. *Each day* within the United States there are close to one million legitimate wire transfers, and internationally about two trillion dollars cross invisible inter-national boundaries. Such transactions can provide convenient masks for illicit transfers. Routine screening or even spot checks of wire transfers must clear high hurdles because most transfers flow through fully automated systems with little or no human intervention. While there have been proposals for using artificial intelligence approaches to identify suspicious transfers, most experts conclude that even with the most advanced software anticipated in the near future, routine monitoring of transfers is not practical.[24]

The simplest transfers available for money laundering are exchanges between accounts in the same bank. Wire transfers between banks with formal, or "correspondent," relationships are also direct and simple. Even when distant banks do not have a correspondent relationship, the transfers from one to the other rapidly transit several intermediate steps, with speed rather than legitimacy the measure of success.

Usually, wire transfers include the following data:

- amount of the transfer
- date of the transfer
- name of sender
- routing number of originating bank
- identification of the receiver of the funds
- routing number of the recipient bank[25]

However, even some of this very limited information may be lost within the various steps of the transfer. The originating bank may simply identify the sender as a good customer. Intermediate banks may drop the routing number of the originating bank. Thus the footprints marking the financial trail are further smudged and difficult to identify.

Bank policies that dictate knowing the customer and reporting to government authorities large or suspicious transfers are difficult to implement in a wired world. As more customers have on-line access to financial institutions and as voice mail replaces telephone conversations, the opportunities to know the customer decrease. The tremendous growth of international trade and business transactions also results in ever-increasing automation with less interaction with customers. In addition, the increasing interdependence of banks worldwide, and particularly correspondent accounts that can be accessed by distant customers, works against both of the preferred policies.

Finally, many nonbanking institutions with no mandatory reporting requirements are now involved in transfers of funds, as well as in barter arrangements. Their legitimate lines of business are frequently used as covers for illegal transfers, particularly in locations where it is necessary to service large immigrant populations. If transfers of ownership of real estate or even of securities replace payments in currency, the trails become impossible to follow.[26]

In short, well-financed and technologically adept criminals are using myriad avenues to subvert the financial systems that are the pillars of international commerce.

Creative crooks are not invincible. Current methods of apprehension are sometimes successful, and more aggressive interventions are being devised. In any event, a good understanding of the innovative schemes to circumvent the law is essential in beginning to choke off the money that feeds terrorism, which brings us back to the Caribbean.

For those who prefer one-stop shopping—to cloak the identity of not only the assets but of themselves—interested parties can respond to advertisements such as the following, placed in *The Economist:*

> We can offer completely legal recognized citizenship, naturalization, including travel documents, driver's license, ID card from different countries which are UNO (United Nations Organization) members. Prices start from U.S. $17,900 for a computer-registered passport from the Dominican Republic up to U.S. $19,900 for a Panama citizenship.[27]

As to the Cayman Islands, it is ironic that at the very site of so many illicit transactions, the government of the Cayman Islands has led the way in the Caribbean region in adopting broad-scope anti-money laundering laws, including mandatory disclosure of suspicious transactions. In 1990, under pressure from the U.S. government, bankers in the Caymans prepared a code of conduct: they would refuse any suspicious cash deposits in excess of $10 thousand. This code did not apply to unsuspicious deposits or to suspect deposits of less than $10 thousand. Also, while the bankers pledged cooperation with American authorities, they recognized that putting an end to hiding assets would be economic suicide. Further, they knew that introducing transparency into their institutions would not solve the drug problem since the laundrymen would simply find other offshore havens.[28]

Today the Cayman government boasts about its enactment of a comprehensive money laundering law in 1996. Its promises to constrain international criminals, however, have yet to materialize, while economic prosperity continues on the islands. With lax enforcement of laws and easy access to its banks, the Cayman Islands are still considered by American law enforcement officials to be one of the most attractive of the Caribbean territories for anyone wanting to launder illicit money from drug trafficking or other criminal activities.[29]

Meanwhile, the banking institutions of the Caymans have become major players in the American mutual funds industry. At the beginning of 1997, more than one hundred mutual fund administrators licensed by the Caymans controlled 1,300 funds with assets of more than $100 billion. Increased international recognition of such legitimate activities in Georgetown helps counter pressures from abroad to spotlight and clamp down on inappropriate dealings which are less visible.[30]

At the core of money laundering operations are the "anonymous" corporations that can be easily established and disbanded in Panama, Liechtenstein, throughout the Caribbean, and elsewhere. The Caymans are the address of record for twenty-six thousand corporations; there are more than one hundred thousand others in the British Virgin Islands and nearby islands. In the Cayman Islands a corporation can be established and maintained for one year for $4 thousand. After handling a single large transaction, such a low-cost facility can be disbanded, complicating the job of even the most skilled investigators.[31]

While the laundrymen are at work around the globe, the perceived failure of the British government to control the tax havens in many Commonwealth countries in the Caribbean (the Caymans are a British Dependent Territory) seems inexcusable. Experts have accused the British foreign service of protecting secret operations that shroud illegal transactions; and the bankers of the

Caribbean where the flag of the British Commonwealth flies know that they will enjoy considerable stability.[32]

In sum, the tax-free black holes in the offshore enclaves become deeper and deeper as more and more simple tax evaders are joined by cadres of criminals perpetrating illegal operations. Both groups use the convenient laundromats of the island states to legitimate their evasion schemes.[33]

Complicating law enforcement efforts is the ease of putting U.S. cash into offshore banks. Panama does not have its own currency and simply allows anyone to deposit American dollars in the banks. On the other side of the world, hundreds of millions of illegally obtained dollars are being transferred to Russia. There they are deposited in exchange for natural resources and other commodities which are sold in Western Europe, where assets suddenly acquire a legitimate money trail.

An unusual banking operation appeared in 1994 when the European Union Bank (EUB) opened for business. "Sited" in Antigua, it advertised itself as the first full-service Internet bank. The bank's charter from the government was conditioned on the bank doing no business in Antigua. Thus, the bank claimed it was exempt from regulation and not subject to inspection.[34]

What were the characteristics of this institution in what has been described as a "banking and Internet free trade zone?" Anonymity was guaranteed. Transactions were not traceable. Patrons did not show up in person. Clients could open accounts, write checks, apply for credit cards, and transfer money around the world with Antiguan law making it a crime for authorities to divulge the identities of bank clients.[35]

EUB allegedly operated in a small room above an Antiguan bar. Other banks in Antigua are known to operate with a single computer and a telephone. One person operates both, simply moving money from one transaction to the next. Of special interest in this case was the revelation that the bank's computer server was not in Antigua after all, but in Washington, D.C., while the principal computer operator was in Canada.

In August 1997, the bank's Web site disappeared. The bank simply vanished. The owners had bilked depositors of millions of dollars with no way to recover losses. Also, the bank is believed to

have laundered large sums of illicit proceeds from Russian organized crime groups.

This bank was one of eleven Russian-controlled banks recently established in Antigua. The government had refused to
shut them down despite the protests of American and British investigators. Antiguan officials allowed senior Russian bank officials to leave the island even after American officials alerted them
to the fraud.[36]

Other distant outposts are learning from the Caribbean experience, and they are stretching even further the tents of secrecy
pitched over shady financial transactions. For example, in 1997
the government of the Seychelles in the Indian Ocean passed an
economic development act ensuring that no questions would be
asked about deposits of $10 million or *more*. What more enticing
invitation could criminals receive? However, after intense international pressure, the government revoked the law.[37]

No discussion of money manipulation could omit recurring
questions about the integrity of our own nation's leaders. High-
profile scandals from Watergate-related money diversions to the
complexities of the Iran-Contra scheme—and going still further
downhill with illegal election campaign financing—send many
signals that higher political purposes can excuse money misuse.
This is precisely the message terrorists repeat to justify their financial sleight of hand.

Richard Nixon reportedly declared, "When the President does
it, that means it is not illegal."[38] According to his logic he bore no
responsibility for the $89 thousand donated to the Committee to
Re-elect the President (CREEP) that was washed through the efforts of a Mexico City lawyer and ended up in the Florida bank account of one of the Watergate burglars. Indeed, Nixon proclaimed
his connections in Mexico were not subject to audit. Nor would he
assume responsibility for the much larger slush funds of CREEP
with purposes which only Nixon considered noble.[39]

John Dean, Nixon's legal adviser, reportedly warned the
President:

> People around here are not pros at this sort of thing. This is the
> sort of thing Mafia people can do—washing money, getting
> clean money, and things like that. We just don't know about do-

ing those things because we are not criminals and not used to dealing in that business.[40]

Ironically, Nixon had little need for the money. He was so far ahead in the polls that his victory over George McGovern was preordained months before election day.

Then, Oliver North set a new standard for White House complicity in the underground financial networks of the world. Despite the subsequent investigations, the facts surrounding the North affair remain hazy. Apparently, North, supported by his superiors at the National Security Council, first convinced the Israelis to sell five hundred American-made TOW antitank missiles to Iran. He eventually increased the number to two thousand. The Iranians paid a Saudi arms dealer Adnan Khashoggi $30 million for the missiles. Khashoggi took his cut and passed the rest to the Israelis who in turn paid North. North arranged a payment to the CIA of $12 million for the missiles and used the difference to finance operations in Central America.[41]

North's laundry cycle reportedly spun money from a Swiss bank account of a dummy Panamanian company to the account of a Swiss accounting firm in the Cayman Islands to the account of the firm's affiliate in Bermuda to the coffers of a Panamanian corporation. The final deposit was made to the account of a shell company registered in Panama and owned by a transportation company that delivered the goods to the Contras.[42] From the outset, North relied on advice from the Bank for Credit and Commerce International (BCCI) which had pioneered the laundering and moving of money not only for the CIA, but also for the PLO, for terrorist Abu Nidal, and even for Saddam Hussein.[43]

Unfortunately, financial scandals involving American politicians, business leaders, government officials, and administrators of nonprofit organizations have become common in recent years. One of the many consequences is the increased difficulty in justifying indignation over international financial scams when some American leaders are equally tarnished.

For ordinary Americans, the Internal Revenue Service (IRS) is supposed to be the watchdog. However, when clearing passenger check points to board an airplane leaving the United States for a foreign destination, have you ever been asked, "Do you have more than $10 thousand in your possession to declare?" The law requires that passengers departing the United States for foreign countries and carrying more than $10 thousand fill out a form for the IRS. Yet, in my many dozens of international departures during the past decade, I have never encountered such a query, although when re-entering the country an analogous declaration is routinely required.

Most honest travelers probably carry credit cards, travelers' checks, and relatively small amounts of cash when going abroad.

10. U.S. Customs Service uncovers currency inside the panel of a van during inbound inspection at U.S. border.

Clearly, the less-than-straightforward travelers planning to leave the country, with their small fortunes hidden in clothes and luggage, seemingly have little to fear. Customs officials rarely ask them, "Show me the money." An Arizona official has estimated that $3 billion per year is smuggled into Mexico, primarily by cars and trucks, across the borders of his state alone.[44]

The Customs Service attributes this vulnerability to a preoccupation with "inbound interdiction efforts." Outbound inspections of travelers are clearly the exception rather than the rule, although there have been occasional enforcement campaigns which resulted in seizures of substantial quantities of cash heading across the border. (See Table 15.)

While individual seizures of cash hidden in body cavities and stashed in personal luggage may seem dramatic, the amount of money is very small compared with the total amount illegally transferred or smuggled out of the United States. There is no

TABLE 15. Examples of Currency Seizures in the Early 1990s

- Individuals have been caught walking across the California border into Mexico carrying shopping bags filled with as much as $500,000 in currency.
- The Customs Service intercepted a woman at JFK Airport who had swallowed fifteen condoms, which contained $7,500, hidden seven additional condoms, containing $3,500, in a body cavity, and concealed $47,894 in her baggage.
- During an x-ray examination of a briefcase, a security guard at O'Hare Airport observed a large amount of currency hidden under a false bottom. The passenger, en route to Paris with a possible connection to Hong Kong, was given three opportunities to declare the money which he repeatedly underdeclared, and $184,200 was seized.
- At JFK Airport, a flight attendant servicing a commercial flight to Colombia was informed of the reporting requirement, and she promptly declared $1,000. However, upon search a customs inspector found $2,600 in a roll of toilet paper, $12,400 in an envelope, $20,000 in a wooden box, $1,627 in her wallet, $5,000 in a carry-on garment bag, $40,000 in a box of laundry detergent, and $13,300 in her jacket pocket.
- Customs inspectors at the Newark seaport seized $7,175,161 from behind false walls in two 20-foot containers loaded with dried peas on a vessel destined for Colombia. The packing and shipping company stated this money, plus $4 million seized at the residence of the shipper's representative, represented drug profits for one month.
- When checking three packages carrying "business documents" a customs inspector at an international courier hub in Memphis found $30,000 in money orders, $30,000 in traveler's checks, and an additional $16,800 in money orders.

Source: *Money Laundering*, General Accounting Office, GAO/GGD-94-73, March 9, 1994.

doubt that a persistent flow of unreported dirty money shipped out of the country will continue, in vehicles crossing the Mexican and Canadian borders, in cargo holds of ships, and on commercial airplanes. Airplanes, obviously, are particularly appealing to smugglers because the travel time is short, the individual smuggler stays close to the money, little preplanning is required, and many destinations are readily accessible.

Like many Americans, I have encountered unsettling financial schemes. While I have not been the target of terrorists nor organized crime groups, my experiences nevertheless suggest that we had better all be on guard as the combination of financial greed and violence becomes ever more widespread.

Following are the details of several incidents in Russia and Nigeria, countries notorious for shady financial dealings:

(1) In 1993, while residing in Moscow, I quickly learned that American dollars—stashed in a money belt or stored in a secured safe— were the most important assets for locals and foreigners alike. Despite repeated claims by the Russian authorities that the ruble was the currency for all seasons, such seasons were short-lived as inflation raged out of control. During periods when dollars were legal tender, certain restrictions applied. For example, Russian exchange offices would accept only bills printed after 1991.

My first financial surprise came in 1993 shortly after I offered a loan of $700 to a Russian colleague for his daughter's airfare to the United States after she won a scholarship to an American college. I obtained 500,000 rubles from a currency exchange kiosk on a Friday. My friend went off with the stack of bills to purchase the Aeroflot ticket only to find the Aeroflot office had closed early that day with no warning or explanation.

As fate would have it, on Saturday the Finance Ministry declared as invalid all ruble notes in denominations of 5,000 or greater that were over two years old. Subsequent news reports claimed the idea was to disrupt the illegal trafficking in rubles in the new Central Asian states of the former Soviet Union. The last thing on my mind, of course, was trafficking in rubles. In fact, astute foreigners and Russians alike rarely owned more than $50 worth of rubles at one time, given the steady inflation rate. Suddenly I was the proud owner of "old" rubles, declared null and void, formerly worth $700.

I wasn't the only one trapped with worthless currency. The protests from influential Russians (undoubtedly a number of crime figures among them) soon moved President Yeltsin, who disavowed any prior knowledge of the plan, to reverse the decision, thereby giving the Central Asians time to dispose of illegally obtained rubles. Within two months, all old bills had been honored at banks, where waiting in line had become a full-day chore. The illegal traffickers didn't miss a beat, and whatever confidence the ruble might have commanded in the West promptly disappeared.

(2) The very week after the declaration that old bills were invalid, I met with the managing director of a major Russian bank to develop procedures for transferring dollars to Russian scientists working on projects funded by the center I headed at the time. My primary concern was whether the bank would have sufficient dollars in its vault to make on-time salary payments. The director assured me that dollars would be there. If there was a run on foreign currency, he would simply telephone a correspondent bank in Montreal, Canada, where a courier would collect $1 million or whatever was needed, board a commercial airliner, and be in Moscow with a briefcase full of $100 bills the next day. When I asked about airport clearance procedures, he replied with the classic Russian phrase "no problem." Although the director's transaction was small peanuts compared to the many other suitcases of $100 bills carried in the holds of many planes flying into Russia— and waved through customs check points in Moscow—it was an indication of the financial porosity of Russian borders.

(3) Then in 1997, a Russian acquaintance told me about another aspect of trafficking in dollars. A Cyprus bank had taken custody of several apartments in his neighborhood. I was familiar with the building where he served as a repairman. I was well aware that the apartments had almost all been leased to joint venture companies as offices. This particular one, the Cypriot bank outpost, was described by my friend as a series of smoke-filled rooms, where ashtrays overflowed with cigarette butts and drawers were stuffed with $100 bills. Every day, streams of well-dressed Russians visited

the apartments and carried on hushed conversations. Some picked up large tote bags and others left them behind. Every second week, the Cypriot manager and his young Russian female accountant traveled to Nicosia, allegedly for rest and recreation—with bulging suitcases—returning to Moscow a few days later with much lighter suitcases to continue their routine.

My acquaintance further reported that until one of the steel doors on the occupied floor opened, the entryway appeared quite ordinary. Once inside, it was clear that the large safes, the burly guards, and the conference tables were not the standard accoutrements of ordinary Muscovites or even of the other offices in the building. At the same time, the only indication of a foreign connection was a small stack of letterhead stationery, regularly used to obtain a plethora of official stamps from Moscow financial authorities.

Was this a center of one of the money laundering operations that facilitate capital flight of tens of billions of dollars each year from Russia to Nicosia and other financial capitals? Likely so, although most of the exiting cash was undoubtedly in the form of illegal wire transfers from Moscow banks and tax-free deposits by foreign customers of Russian exports. The Moscow outpost was probably an *expediter* service to encourage many Russian heads to turn the other way when transactions were in process in the financial institutions of Moscow.

(4) Another incident provided personal insights into the well-reported Nigerian financial manipulations that have fleeced many American bank accounts. My wife had just attended the 1996 Annual State Department Security Briefing for Businesses Operating Abroad where Nigeria was fingered as one of the most corrupt nations in the world. A popular method of ripping off funds from foreigners, experts reported, was simply to fax them and ask for use of their bank accounts as clearing houses. Unfortunately, those who volunteered to help "mistreated" Nigerians soon found their bank accounts cleared out completely.

We never expected to obtain any more details than those presented in the briefing. Several months later, however, a friend received a fax from one such Nigerian group. Not knowing that the State Department was already on the case, she thought I could bring the scam to the attention of government contacts. The letter read:

> First, I must solicit your strictest confidence in this transmission. This is by virtue of its nature as being utterly confidential and "Top Secret." You have been recommended by an associate who assured me in confidence of your ability and reliability to prosecute a pending business transaction of great magnitude, requiring maximum confidence.[45]

The letter of October 21, 1996, signed by Dr. Abe Odum in Lagos, Nigeria, supplied details which apparently had hooked Americans to give away bank account numbers that were promptly exploited by the Nigerian cyberbandits.

According to the letter writer, he and a few other civil servants had access to $31,320,000 "floating in the Central Bank of Nigeria." They could arrange for the transfer of the funds into the bank accounts of willing American accomplices. The account owner (identified in the letter as "you") was to receive 20 percent. "They" were to receive 70 percent. Ten percent was to be used to cover expenses. As an added enticement, you were to receive "at less than market price as much as 500,000 barrels of Automatic Gas Oil for spot lift."[46]

All he needed to trigger these transactions were your company's name and address, your banker's name, and your telephone/fax number. This information was to "enable us to write letters of claim and job description respectively enabling us to use your company name and account details to apply for the payment." The faxed letter was so poorly written it was difficult to imagine many takers. According to television reports documenting the scam, there have been a surprising number of victims of these types of Nigerian invitations.

Threatened with law suits, the Central Bank of Nigeria has responded with large ads in the American press warning "all and sundry" about the modus operandi of the international syndicates the bank called a source of "embarrassment." The ads have stressed, "You are warned and advised in your own interest to ignore the get-rich-quick business solicitations." The ads urge that solicitations be reported to local law enforcement agencies or to Interpol.[47]

(5) Finally, on several occasions, I have personally witnessed crime and intimidation that seems to be the rule in Nigeria. At the Lagos airport I have seen foreigners, whose yellow inoculation cards were judged incomplete, required to pay $100 or be subject to puncture by a syringe filled with an unknown serum. Exiting the airport, the traveler must dodge a second self-appointed customs service that tries to again inspect baggage, or more accurately, steal baggage. Also, I was in Lagos on a day of public executions for crimes against the state, another chapter in the battle over political governance that has raged for many years.

Now, organized Nigerian criminal groups reportedly transship 40 percent of the heroin reaching the United States, and the country's reputation of excellence in forging documents continues to grow. To prove their commitment to eradicating crime, Nigerian authorities once announced they had arrested a criminal at the Lagos airport who had $800 million in cash. Their boast soon lost credibility once you figured out that if the booty was in $100 bills, the suspect would have been carrying several tons of cash.[48]

In mounting their counteroffensive to the expanding money laundering operations around the world, representatives of all concerned U.S. law enforcement agencies spend about two months at the beginning of each year jointly assessing the financial vulnerability of each of 200 nations and territories. They rank the jurisdictions as to the priority they should receive in U.S. efforts to combat illicit financial operations. As a point of departure, the agencies have prepared a checklist of the characteristics of a country that would guide crafty money managers to seek it out as a haven more hospitable than others for their dealings. (See Table 16.)

With these and related considerations in mind, the U.S. government has identified many of the familiar players as high-priority countries requiring urgent attention: Aruba and the Netherlands Antilles, Cayman Islands, Colombia, Cyprus, Hong Kong, Italy, Mexico, Nigeria, Panama, Russia, Singapore, Thailand, Turkey, and Venezuela.[49]

Many Americans may be somewhat surprised to see other countries on the high-priority list. Canada is an attractive venue

TABLE 16.   Characteristics of User-Friendly Countries for Money Laundering

- Failure to consider money laundering as a serious crime, or specifying that laundering is a serious crime only if associated with drug trafficking or other narrowly designated activities.
- Rigid bank secrecy that cannot be penetrated for authorized law enforcement investigations.
- Minimal identification requirements to conduct financial transactions, or widespread or protected use of anonymous, nominee, numbered, and trustee accounts. No required disclosure of the owner of an account or the true beneficiary of a transaction.
- Lack of effective monitoring of currency movements.
- No mandatory requirements for large cash transactions.
- No mandatory requirement for reporting suspicious transactions, or pattern of inconsistent reporting under a voluntary system, or lack of uniform guidelines to identify suspicious transactions.
- Use of monetary instruments payable to bearers.
- Well-established nonbank financial systems, especially where regulation and monitoring are lax.
- Patterns of evasion of exchange controls by nominally legitimate businesses.
- Ease of incorporation, especially ownership that can be held through nominees or bearer shares, or where off-the-shelf corporations can be acquired.
- Limited or weak bank regulatory controls, especially in countries where the monetary or bank supervisory authority is understaffed, underskilled, or undercommitted.
- Well-established offshore or tax-haven banking systems, especially countries where such banks and accounts can be readily established with minimal background investigations.
- Extensive foreign banking operations, especially where there is significant wire transfer activity, multiple branches of foreign banks, or limited audit authority over foreign institutions.
- Limited asset seizure or confiscation capability.
- Limited narcotics and money laundering enforcement and investigative capabilities.
- Countries with free-trade zones where there is little government presence or other oversight authority.
- Patterns of official corruption or a laissez-faire attitude toward the business and banking communities.
- Countries where the dollar is readily acceptable, especially countries where banks and other financial institutions allow dollar deposits.
- Well-established access to international bullion trading centers in New York, Istanbul, Zurich, Dubai, and Bombay.
- Countries where there is significant trade in or export of gems, particularly diamonds.

Source: "Financial Crimes and Money Laundering," *International Narcotics Control Strategy Report*, 1996, U.S. Department of State, March 1997.

for currency from the United States due to its close proximity, advanced financial sector, and lack of mandatory reporting requirements. In Germany, money laundering goes hand-in-hand with smuggling of all types, gambling, and sex-slave and narcotics traffic. Laundering thrives in Dutch money exchange houses, casinos, credit card companies, and insurance and securities firms. Switzerland has significantly changed its previous patterns of secrecy of late. Nevertheless, it remains a tax haven, and to a lesser extent a money laundering site, with no controls on transborder currency flows and uneven enforcement throughout its cantons. Finally, the United Kingdom has become subject to criticism due to its somewhat indiscriminate support of British international financial institutions, as we have seen.[50]

Meanwhile, the U.S. Department of State continues to focus on Latin America as a whole, where drugs, crime, money laundering, and corruption are becoming intertwined among most of its countries. In early 1996, twenty-one Latin American countries signed an international agreement against corruption. The basic thrusts are to require national antigraft laws and to promote international cooperation in tracking down wrongdoers. Under the new convention, countries must now provide details of bank accounts under their jurisdictions if another country asks for them. But extradition remains dependent on the laws of the countries or on existing bilateral treaties. With Costa Rica openly advertising that it is a country of asylum, major escape routes clearly remain.[51]

In a related activity to clamp down on illegal financial dealings, in 1996 the countries of the region adopted *Model Regulations on Money Laundering* and began regular consultations on the process of money laundering and methods to curtail it.[52]

Despite a growing list of interventions, U.S. investigators following the trails of financial fraud frequently become discouraged as the drug cartels remain one step ahead of them, hop-scotching around the world to find new sanctuaries for their funds as old repositories are placed off limits. Their frustrations are becoming of much greater concern as the criminals buy equity in international banks, as we have seen, and these banks provide even safer havens.[53]

Also, while American authorities widely publicize drug seizures and associated shipments of cash, the traffickers frequently scoff and view such developments as merely the cost of doing business. They quickly replenish their drug pipelines, sometimes even buying back from corrupt middlemen drug cargoes previously seized in the Caribbean.[54]

The uneven progress in reducing demand for narcotics in the United States and the uncertainties in interdicting supplies lead to a gloomy outlook for winning the drug war. At the same time, the five-fold increase in the cost of laundering drug money indicates that the pressure is on, but we are still a long way from toppling the financial infrastructure that supports the traffickers.

We should not lose sight of the implications of all this cash movement as it relates to incipient terrorist activity. With the growing costs of technical arsenals preferred by today's terrorists, we must assume that they too will rely on standard schemes to shelter funds from taxation while incorporating elaborate approaches of the drug traffickers when necessary. Meanwhile, a number of governments are increasingly willing to impose new restrictions on drug-related financial crimes, and they must now extend such measures to all sources of illegal money that feed terrorism.

The general strategy for engaging other countries in a worldwide assault on money laundering seems clear. Exchanges of many types of intelligence and financial information should be expanded. Nations should regulate the financial transactions of not only banks but other financial services institutions. Finally, stronger enforcement of regulations is critical if we are to stop the dangerous flow.

Of greatest importance, government officials around the world now accept in principle the premise that financial institutions must know their customers. Without such an operating principle, money launderers will run free. At present, the ease of placing currency into an account for immediate transfer to other jurisdictions is the weak link in attempting to prevent laundering.

As regulations increase and limitations tighten, progress can be made in curbing money laundering. Those participants in the process need to be convinced that when caught, they will go to prison, their funds will be seized and their accounts frozen. When progress is made in that direction, the terrorists will also discover that they need new models for financing their operations.

One breakthrough for the law enforcement community was the passage of *The Digital Telephony and Communications Privacy Improvement Act of 1994*. This law requires common carriers (e.g. AT&T, MCI, Sprint) to retrofit their systems to provide law enforcement agencies in possession of appropriate warrants access to their communications and dialing information. However, in softening the request of the administration, Congress decreed that such access is limited only to those communications specified in the warrants and cannot be granted for remote, continuous monitoring in the hopes that incriminating information eventually becomes available. This law does not apply to providers of access to Internet, such as America on Line (AOL), and the issue of legal eavesdropping on Internet is the center of a raging debate as we will see in Chapter 9.[55]

One final incident incorporates many of the described techniques used by organized criminal gangs working across international boundaries.

On June 26th, a bank cashier and a clerk drove through St. Petersburg, Russia, carrying millions of rubles for an interbank transfer, accompanied by two policemen and five soldiers. Ten people were waiting to rob them in Yerevan Square, acting on information from corrupt government officials that the bank transfer was about to occur. The leader of the robbers was dressed in an army uniform. He created turmoil in the square by warning pedestrians that violence was expected. When the bankers arrived, the conspirators detonated explosives and stole the currency. The stolen currency was hidden in a government building in the Caucasus Mountains for a month before it was smuggled across the border and eventually into Germany.

Five months later, the gang began laundering the money by placing it in small accounts as they distributed it into banks across northern Europe on a single day. At that point, the launderers and some of the ring leaders were arrested on information supplied by an undercover officer with the German police, who shared it with the Russian police. This example presents many of the elements of the problem: violent robbery, corruption, smuggling of currency, and money laundering. And, interestingly, the same gang had stolen currency plates and was experimenting with making counterfeit rubles.[56]

These events did not take place in June 1997, but in June 1907. In 1997, the circuitous financial escapades of five months could easily be compressed into five minutes.[57]

Now, as in 1907, the challenges are to toughen anti-laundering laws, improve financial-intelligence gathering, and seek international cooperation. As pieces of plastic bearing microchips are loaded with e-cash and payments over the Internet expand, the technological complexities in fending off fraud increase. Banks can no longer serve as the only choke points for monitoring transactions, and the difficulties in determining jurisdictional authority for pursuing criminals increase. Meanwhile, sophisticated technological countermeasures are being found; already the possibility of attaching unique electronic serial numbers to each e-mail transaction has caused criminals to retain for future use the old standby cash-in-the-bag approach.[58]

The challenge is to inoculate the new electronic systems against crime while still permitting the rapid growth which can benefit economies throughout the world. The formula offered by Senator John Kerry certainly warrants serious consideration:

> Start from the premise that big-league criminals should not be permitted to keep the proceeds of their crimes, regardless of where these proceeds have been invested. Add to that the agreement that ownership of financial assets must be transparent to law enforcement: Every country should create systematic recordkeeping requirements for financial and real property and allow these records to be shared in cases involving serious crime. Add in procedural safeguards to ensure a public hearing process before seized assets can be forfeited to the government. Encourage smaller governments to share information whenever a big-time crook uses their jurisdiction, and provide quick rewards through asset sharing. Take them together, and you go some distance in shifting the balance of forces from the bad guys toward the cops.[59]

Finally, while we have concentrated on how money laundering works to the advantage of independent terrorist groups, we cannot forget the activities of irresponsible states and institutions. Indeed, BCCI was involved with Saddam Hussein and has been linked to the Pakistani nuclear bomb program.[60] In short, *money is the lifeblood of modern terrorism*. The more we can do to reduce the supply, the less the blood will flow.

CHAPTER 8

Business as Usual: Hostages, Bombings, and Assassinations

*Americans do not know what is meant by
airport security.*
Senior Israeli Aviation Official, 1996

*The effects of a physical attack on our energy, banking,
transportation, or communications network could
spread far beyond the radius of the bomb blast. The damage from a cyberattack might not be known for months.*
President's Commission on Critical Infrastructure
Protection, 1997

*A 4.5 mm. single shot reloadable assassination device
firing a cyanide-tipped fragmented projectile hidden in a
common lipstick case.*
Sign over an exhibit of 1952 East European
technology, CIA headquarters, 1997

*T*wo burly female security guards ushered the disheveled
nun through the security checkpoint at the Rio de Janeiro
airport and told her to stand in line behind me. She was clearly
angry as she jerked at her frock in an unsuccessful attempt to adjust its bulk. To underscore her obvious fury, she was protesting
loudly in Portuguese. Her soliloquy was not to be mistaken for
a prayer of forgiveness toward the security police, according to
my interpreter.

After all, the nun had just been strip-searched in a small office adjacent to the waiting room, where the passengers for the flight to Recife in Northeast Brazil were awaiting instructions. She obviously had been "profiled" by Brazilian security as a likely terrorist.

The identification of nuns as potential bombers might normally seem odd, but I remembered the newspaper headlines of the previous morning. The very same flight had been hijacked to Venezuela two days earlier. The leader of the hijackers had been dressed as a nun, wearing a cast on her arm. Once the plane was airborne, her cast came off. She pulled out an automatic pistol concealed in her cast, and commandeered the flight to Caracas.

This actual incident took place in the early 1970s. Today, the same tricks of hiding weapons and contraband in medical casts and of disguising identities with clergical garb are still quite popular among the terrorist crowd.

At the time of the original incident, I had already become accustomed to the hypersensitivities of Americans in Brazil to terrorist acts. Earlier that week, I had been the guest of the American Consul in Puerto Allegro, the principal city in the southern region of the country. The consul was eager for local Brazilian leaders to meet me, a senior official of the U.S. Agency for International Development.

The consul himself collected me at the airport, although his greeting was less than enthusiastic. Several days earlier he had been the target of an assassination attempt. His doctor would not permit him to shake hands or even turn his head. Local terrorists were playing on anti-American sentiment in their efforts to disrupt an election campaign. While a bullet had been successfully extracted from his shoulder, the consul was clearly not at his diplomatic best.

The consul and a fleet of security cars led the way to the official American residence, which was surrounded by heavily armed guards. They had just completed a periodic sweep of the area and had cordoned off the street right in front of the residence lest a terrorist truck deliver a bomb.

Within a few hours, many Brazilian guests had arrived for the consul's reception in my honor. Other than their many queries as to the condition of our wounded host, little was said about the shooting or the unusual security procedures. Apparently, such incidents happened so frequently they were considered the norm for the city at that time.

Now, often reading headlines in our own country such as "Security Breach Brings Dulles Airport to a Halt" and "Authorities Evacuate Washington Square Building for Tenth Time in Three Weeks after Receiving Bomb Threat," we wonder if hijackers and bombers have finally found their permanent niches not only in Brazil and other distant lands, but here in the United States as well.[1]

We have already examined the emerging terrorist threats accompanying the diffusion of weapons of mass destruction and the spread of nuclear, biological, and chemical materials—enhancements that can wreak havoc in even primitive delivery devices. We also have seen how drug dealers, organized crime groups, and money launderers are linked to arms trafficking, an enterprise that stocks weapon arsenals around the world. Here I will elaborate on the continuing terrorist dependence on—and sometimes preference for—what might be described as their traditional technologies. This discussion helps round out the sphere of dangers that confront the United States and the world as we enter the next century.

Terrorists will continue to use pistols and pipe bombs, and they will employ more recent innovations, such as miniaturized guns and plastic explosives. Preventing damage and death from such familiar instruments of destruction—in the air, on land, and at sea—will remain the core of many counterterrorism efforts in the years ahead.

When TWA flight 800 exploded into fragments just off the coast of Long Island in July 1996, heads in America immediately turned toward the Middle East in search of radical Muslim perpetrators. This incident was presumed to be a delayed encore to the Libyan-engineered destruction of PanAm flight 103 over Lockerbie, Scotland, in 1989. While cautious in its initial assessment, the U.S. government nevertheless mounted a massive search for Muslim extremist criminals who might have orchestrated the deaths of 230 Americans. Immediately following the incident, the FBI and other law enforcement agencies theorized a bomb had been planted in the baggage compartment of the plane. The overseas effort to determine responsibility, led by the Department of State, the CIA, and the National Security Agency, included electronic eavesdropping, quizzing dozens of informants, and offers of large cash payments for leads.[2]

Casting a shadow of suspicion was a trial under way in a nearby Federal District Court in Manhattan. At the time of the airline tragedy, the U.S. Department of Justice was presenting its case against Ramzi Ahmed Yousef and two coconspirators accused of plotting to kill up to four thousand Americans traveling in twelve commercial airliners over various Pacific routes. Yousef was also under suspicion as a key organizer of the World Trade Center bombing three years earlier. The plot against jets over the Pacific Ocean was portrayed by the prosecutors as a protest against U.S. support of Israel.

In planning his crime, originating in Asia and hopscotching about the globe, Yousef had masterminded the design of tiny bombs, undetectable to even the most sophisticated airport scanning devices. Liquid explosives were disguised in ordinary contact lens solution bottles, with their timing mechanisms made from Casio digital watches. The idea for one segment of the journey was as follows: A terrorist would fly from Manila to Seoul on a Northwest Airlines flight. En route he would plant in a lifejacket under the seat a bomb that would go off on the jet's next leg to Los Angeles. He would then take a United Airlines flight from Seoul to Taipei and plant a second bomb that would explode as that plane headed for Honolulu. Finally, taking the next United flight to Bangkok, the villain would plant a third bomb to go off on still another plane enroute to San Francisco.

The U.S. government charged that Yousef field-tested the system by planting a bomb with the signature Casio timer on a Philippine airliner headed from the Philippines to Japan. Yousef deplaned at an intermediate stop at the Philippine resort city of Cebu and the bomb went off on the next leg to Tokyo. Fortunately only one Japanese passenger was killed. Yousef was soon captured, and his trial resulted in a one-way trip to jail, with no chance of release.[3]

Despite the political impact of airplane hijackings, seizures of passengers as hostages, and bombings of airports and airplanes, such events no longer dominate the terrorism landscape as they did in the 1970s and 1980s. The combination of tighter airport security, more forceful responses to incidents, and improved international cooperation both in resisting international air piracy and in extraditing perpetrators explain the decline in events. At the same time, many experts predict that the threat of terrorism against the United States will increase and that aviation will remain an attractive target for terrorists.[4] If the smuggling of money and drugs via the airlines is factored into the equation, airline security measures become even more critical than in the past. A brief review of lessons learned during the heyday of airline incidents helps establish the foundation on which to fashion the increased security measures required in today's world.

As we noted previously, many of the early incidents took place in the Middle East where airplanes from many countries fell victim to the seizures. Soon aircraft in Europe, Asia, and North America were included in the target lists, with destruction of the planes often being the terrorists' preferred outcome. (See Table 17.) While the 1990s have seen a decline in incidents, predictions are that airline flights will remain a target for terrorists.[5]

Of the more than nine hundred attempted skyjackings since the 1930s, none have been more intriguing than the attempts featuring parachute escapes (parajacking), particularly in the United States. In 1971, D. B. Cooper bailed out over the state of Washington with $200 thousand he had extorted from an airline after seizing a

TABLE 17.   Significant Periods in Evolution of Skyjacking

- 1947–52 East European seizure of planes to gain political asylum.
- 1958–61 Castro-initiated political skyjacks in Oriente Province to help gain power. Skyjacks by backers of Batista to escape to the United States.
- 1968–72 Reverse flow of homesick Cuban skyjackers back to Cuba.
- 1968–71 U.S. skyjacks by emotionally unbalanced, many to Cuba.
- 1968–73 Arab skyjack war in Jordan, Lebanon, Algeria, and other countries.
- 1971–73 Criminal extortion phase featuring parajacking in the United States.
- 1973–78 Western response with airport security, bilateral agreements, international conventions, and counterterrorism teams.
- 1974–78 Continuation of low-key Arab skyjack war against western airlines.
- 1983–88 Moslem fundamentalist expansion of Arab skyjack war; Sikh skyjack activity.
- 1998–96 Decline in aviation incidents which are no longer related to a pattern.

Source: Peter St. John, *Air Piracy, Airport Security, and International Terrorism*, Quorum Books, Westport, Connecticut, 1991, p. 6; "Criminal Acts Against Civil Aviation, 1995," Office of Civil Aviation Security, Federal Aviation Administration, 1996.

plane at the Seattle airport. During the next seven months, five other Americans followed in his slipstream carrying a total of $1.5 million in extorted funds. Unlike Cooper, whose fate remains unknown, all were captured, and 80 percent of the funds were recovered.[6]

The popularity of parajackings quickly grew, and a cult interested in this type of escapade rapidly emerged. Fourteen additional heists were attempted in the United States, but none were successful. Overall, the twenty attempted skyjacks racked up ransoms of about $9 million, with about $7 million recovered.[7] While never seriously used by terrorists, parajacking added a sense of daring and bravado to efforts at outwitting the system—a perpetual motivator of some criminal behavior.

Over the years, a favorite subject of researchers has been the characteristics of perpetrators of aviation incidents. As previously described, terrorists are quite normal people, following logical pathways and focused on well-defined objectives. The preponderance of today's terrorists probably will not find the limited space of an airplane to be the ideal environment for making their point and then getting away with it. Nevertheless, in trying to anticipate future incidents, airline leadership clearly pays attention to the historical record of the earlier terrorists, sometimes classified as the three Cs: crazies, criminals, and crusaders.

Crazies who have stormed aircraft have been described as sacrificial and sometimes delusional. They are usually incoherent and ready to plea for personal help as their psychodramas approach their finales. Generally loners eager to connect with their audiences, they have been amateurish in their tactics and ready to take high risks. They can be persuaded but not bought. They may be receptive to moral appeals and will pay more attention to offers of personal protection than to threats of punishment.

Criminals, on the other hand, assiduously avoid personal interactions with hostages and are interested in concrete outcomes of their actions. They are professional and take few risks. They can be bought, but not persuaded. They respond to specific business-like deals. They are ruthless when necessary to achieve their immediate goals.

Finally, crusaders are publicity-oriented, seeking attention both within the airplane and outside. They are often suicidal and/or homicidal. They seize hostages for barter or for brainwashing.

They are often violent, and they seek deals on their own terms. They will consider specific settlements of symbolic value and welcome opportunities to make vague moral appeals. They cannot be corrupted into abandoning their announced missions.[8]

That said, strategies to cope with modern terrorism in airports, airplanes, and elsewhere must be designed to deal with both normal and bizarre behavior, both individual perpetrators and representatives of criminal organizations, and both moderate and outrageous demands.

A specific example helps tie concepts to reality. In 1986, the Syrian-sponsored scoundrel Nezar Hindawi impregnated an Irish chambermaid in England and promised to marry her in Israel. He booked her on an El Al flight and told her that as an Arab he would have to fly on a different airline and then join her in Israel.

In the taxi on the way to Heathrow airport, Hindawi asked his girlfriend to take a traveling case with her. It contained a cleverly designed bomb that did not raise alarms during the airport baggage x-ray screening. However, an El Al guard at the boarding gate eyed the traveling case with suspicion and searched it. There was a flat plastic explosive charge in a false bottom and an ordinary calculator equipped with an extra electronic circuit to fire the bomb. It was to be detonated at high altitude, bringing down the plane and all passengers including the unsuspecting girlfriend and Hindawi's own unborn child.[9]

Was Hindawi a crazy, a criminal, a crusader, or simply a reluctant groom? Chances are he was a little bit of each.

The number of airline travelers will continue to rise well beyond the recent level of 1.5 million passengers per day in the United States alone.[10]

Security interrogations and personal scrutiny of travelers are likely to increase, particularly in the United States. Currently such inquisitions last less than one minute. The questions as to who packed one's luggage, where it has been stored since being packed, and whether it includes packages from others are usually delivered in a perfunctory manner by personnel numbed by repetitive queries and innocuous answers.

El Al airlines, a perpetual terrorist target, has for many years given as much attention to the careful identification and screening of passengers as to the detection of dangerous objects that could be slipped through physical barriers. Long ago, El Al developed five types of psychological profiles as the basis for sophisticated passenger interrogation techniques. (See Table 18.)

If El Al personnel have lingering uncertainties about specific passengers, they may telephone families, acquaintances, or police to resolve suspicions before clearing entry into boarding areas. The approach of careful passenger scrutiny, together with rigorous control of all items being loaded, seems to have worked, since there have been no hijackings or bombings of El Al planes since 1970.[11]

The El Al system is time and labor intensive. Passengers are instructed to arrive at the airport two hours or more in advance. While suitable for the small number of international flights of the airline, comparable procedures do not seem feasible for the huge number of flights in the United States. Nevertheless, the El Al experience can provide lessons and approaches to help guide American efforts in targeting those individuals who are most likely to pose a terrorist threat.

In early 1997, a U.S. government panel led by Vice President Al Gore strongly endorsed a system of computerized passenger profiling to help single out those individuals whom certain indicators suggested should have their possessions more carefully scrutinized. Airlines were asked to collect many categories of data for constructing passenger profiles—for example, whether tickets

TABLE 18.   El Al Classification of Skyjackers

Naive: A passenger who carries a bomb in his or her baggage without being aware of it, perhaps thinking it is a gift.

Partly Naive: A passenger who has been tricked to believe that his or her baggage contains contraband, such as drugs or cash, but in reality it conceals a bomb.

Framed Terrorist: A terrorist who has been tricked to carry a bomb without being aware of it, perhaps thinking it was unassembled explosives.

Terrorist Hijacker: A terrorist determined to seize the plane.

Suicide Terrorist: A terrorist prepared to go down with the plane.

Source: Airline Passenger Security Screening, National Research Council, National Academy Press, Washington, 1996, p. 13-21; Karen K. Addis, Profiling for Terrorists, *Security Management*, May 1992, pp. 27-33.

were bought through a travel agent or by credit card (use of cash could be a tip-off of possible trouble); the dates of ticket purchases, (last-minute purchases raise questions); and traveler destinations (rogue states obviously raise eyebrows as well as red flags). Traveling with a child might remove some suspicion, as would using round-trip tickets, and renting a car on arrival. Less reliable characteristics are a person's appearance and mannerisms, details that El Al has always considered essential.[12]

The panel recommended putting in place some limits to information gathering, such as barring the airlines from collecting data on a passenger's race, religion, or nation of origin. It also underscored that searches must not stigmatize any traveler; it added that airline files cannot be permanent, and old information no longer needed must be systematically destroyed.[13]

Arab-American and civil liberty groups immediately lodged protests, arguing that passengers should be required to check their luggage, not their constitutional rights. They contended that even with safeguards, the proposed profiling policy violates the right to privacy and would cause humiliating travel delays for people with dark skins, national dress, and unfamiliar names. They pointed out that airline computer systems obtain data from law enforcement databases that may record a person's arrest but not his acquittal. Also, they were concerned that the policy discriminates against poor people who do not qualify for credit cards.[14]

As long as there are only a few airline incidents, the advocates of civil liberties will undoubtedly find sympathetic audiences for their pleas. With each highly visible incident aimed at American victims—particularly those tragedies shown to have a foreign connection—the traveling public will no doubt demand ever tighter security procedures. Meanwhile, the overwhelming majority of air travelers, preferring to be searched and not scorched in a fiery crash, will accept without complaint whatever security measures adopted.

Closely linked to passenger profiling are the procedures for physical screening of passengers, baggage, and cargo. The airport configuration is of course an essential starting point to ensure that people and contraband do not circumvent these security procedures. One extreme proposal calls for electrified fencing on the outer perimeters of airports, with entry tightly controlled through a few gates. While not in favor of such fortress-like measures, most airport authorities agree that all entries beginning with the general arrival and departure areas need greater attention.[15] With the increased awareness of the possible releases of biological or chemical agents, air conditioning and heating systems have become focal points of attention for security personnel when designing, retrofitting, and policing airports.

Typically, passengers place their carry-on luggage on conveyor belts for x-ray inspections and then walk through portals that detect metallic items. If suspicious items are exposed by the x-ray screening or if alarms sound at the metal detector, more detailed searches are undertaken for dangerous items. While these techniques are very well developed, the poorly paid security personnel who apply them are often less than fully attentive. The need for upgrading the performance of such personnel is a widespread concern.

El Al has adopted a further precaution of requiring that all carry-on baggage be stored in a compartment in the front of the plane, thus reducing the opportunities for using weapons that are smuggled aboard in such baggage.[16]

Turning to the screening of checked luggage, detection devices currently in use, primarily x-ray equipment, have a number of limitations. They vary in their ability to detect the types, quantities, and shapes of explosives. They frequently trigger false alarms, daily requiring hundreds or even thousands of individual bag searches in a busy airport. They depend on security personnel to resolve alarms, through closer examination of a computer image or a hand search of the item in question. The quality of the human judgments as to whether a dark image signals trouble is often questionable.[17]

The standard procedures are not adequate for detecting nonmetallic weapons, plastic explosives, or wrist-watch bombs. New technologies are evolving to address such inadequacies—including millimeter wave imaging and x-ray imaging to expose items hidden under clothing and in luggage and mechanical sniffers for detecting chemical vapors emanating from travelers or baggage. As these technologies mature and become available for deployment, new questions will arise over cost, passenger acceptance, and air carrier and airport implementation of procedures. Also of concern are issues of health, typified by perceived risks associated with exposure to x-rays. The invasion of privacy, including the use of technologies that display images of the unclothed body, will of course persist. Passengers might be given the choice of subjecting themselves to alternative approaches but such options inevitably would lead to greater delays.[18]

Under development are advanced x-ray machines that have lower detection capabilities but can process baggage much faster. Also, tomography devices, based on advances in the medical field, provide three-dimensional images: they have very high detection capabilities but low processing speeds. The attributes of these two technologies need to be married to allow both faster processing and higher detection. In addition, electromagnetic radiation devices that detect specific explosives are in the developmental pipeline. Within the next decade, technical advances will greatly enhance screening procedures but at a substantial cost.[19]

Security problems are even more difficult when handling commercial cargo and mail, in view of the common practice of batching large shipments. The devices used for screening checked baggage do have some applicability to cargo. In addition, very sophisticated nuclear-based technologies which generate gamma rays and neutrons are under development, although their use will require considerable shielding of nearby security personnel.

Another trend is the use of blast-resistant containers for cargo or baggage. Such containers are strong enough to contain the explosion and fragments from an in-flight detonation of a small explosive device. Of course, such containers require space; their weight will likely displace a few passengers. Thus, the direct and indirect costs will be substantial. A related approach is the Israeli system of subjecting all luggage to a depressurization test before loading, since bombs activated by pressure changes have at times been the weapons of choice.[20]

Often the most effective bomb detector is the trained dog, with a snout ten thousand times more sensitive than the human nose and much more reliable than electronic sniffing devices. The Gore Commission recommended prompt deployment at airports of 114 additional teams of dogs capable of sniffing out explosive devices.[21] However, this proposal quickly bogged down as six government agencies squabbled over canine sniffing standards.

For example, the Federal Aviation Administration (FAA) objected the Bureau of Alcohol, Tobacco, and Firearms' requirement that to be certified, a dog must have a 100 percent performance record in detecting explosives in sixty disguised containers. The Department of Defense and Secret Service also dislike the test, but they can't supply a better alternative. The basis for the objections—too stringent a test, wrong types of containers, or perhaps simply a disagreement over which agency should develop the test—has not been reported.[22] Nevertheless, while researchers are baffled as to just what odor or chemical stimulants trigger a dog's responses, many government agencies already rely on canine olfaction as the trump card in countering smugglers of explosives, drugs, and chemical agents.[23]

Most of the research concerning detection technology focuses on physical detectors of vapors, enhanced x-rays, and neutron detection devices. Still, the eyes, ears, noses, fingers, and voices of experience leading to suspicions and hunches remain critical components of the detection armory. The best security comes from using a variet detection methods—physical, canine, and human—each searching for a different characteristic be it a metallic component, vapor, a nervous twitch or suspicous molecules.[24]

Finally, screening procedures for maintenance workers, service personnel, and others with access to the airplanes should be no less demanding than those confronting passengers. Even extraordinary services, such as last minute sprints to load cargo-containers of human organs being rushed to transplant patients, must go through appropriate security checks on the way to cockpits or aircraft storage bins. Then, too, diplomats, while placed at the head of the line, should have no right to claim exemptions from search when security is at stake.

Whether planning violence in the air or on the ground, terrorists like plastic explosives since they are malleable, lightweight, and stable. They can be molded to fit suitcases and shaped into toys. They are not easily spotted by x-ray machines or metal detectors, and less than one pound of the substance can blow a hole in the side of an airplane.

Semtex is a favorite plastic explosive. Once a major export of Czechoslovakia, Semtex now seems to be more carefully controlled than in the past when rogue governments and unidentified groups were among the best customers. *Tons* are believed to be available in countries such as Libya.[25] Semtex isn't the only plastic explosive that may be included in enemy inventories. Concerns are increasing that C-4, produced for the U.S. Army, has also made its way to the international black market. Meanwhile, instructions as to how to make plastic explosives are posted openly on the Internet. While more complicated than fertilizer and black powder bombs, plastic explosives can be produced by many chemists who have advanced training, access to modest laboratories, and a few readily available ingredients.

An international treaty calls for nations with stockpiles of plastic explosives to use or destroy them within specific time limits. Also, the treaty requires that all future plastic explosives be chemically marked so that they can be more easily uncovered by airport bomb detectors. The treaty has been ratified by a number of nations, including the United States, but more signatures are needed before it enters into force.[26]

For many years, law enforcement authorities have argued for the chemical marking of not just plastic explosives but *all* explosives. If chemical marking was used, bomb experts would be able to sift through the rubble of a crime scene and find the markers that in turn would assist in determining the origin of the explosives.

However, the efficacy of the technologies for marking or "tagging" the more conventional types of explosives has been hotly debated. Proponents of tagging schemes include the manufacturers of taggants who claim they can safely tag anything from high explosives to fertilizers to gasoline—all at very low costs. On the other side of the controversy are the manufacturers of the bomb-making materials. They argue that if the taggant proposals were adopted, they would be swamped with paperwork for tracking the movement of millions of tons of chemicals. They refer to studies showing that taggants may introduce instabilities into otherwise safe chemicals that might cause premature explosions. Still, the advocates of tagging contend that taggants are perfectly safe, pointing to Switzerland's twelve years of safe experience. Taggants are required for all explosive materials produced in that country. Also, the advocates are quick to add that a taggant helped in tracking down and convicting a bomber in Baltimore.

While taggants help investigators find criminals, law enforcement officers need to be able to prevent the bombings in the first place.[27]

Dramatic video footage of bombed-out buildings and emergency teams crawling through rubble in search of bloodied victims have displaced photos of aviation standoffs as the manifestation of the most immediate terrorist menace.

Few countries can compete with the United Kingdom in bomb attacks, at least until the recent agreement among the eight parties that have been involved in the hostilities in Northern Ireland. Both in Northern Ireland and in downtown London, the frequency of IRA bombings had numbed people to the dangers of such incidents. For Americans, these bombings had become a part of the accepted image of Great Britain—frequent explosions unique to that country's political problems.

There were notable exceptions to the relative lack of American concern over developments in London, Belfast, and Dublin. In 1992 and 1993, the IRA exploded huge truck bombs containing several thousand pounds of improvised explosives in the heart of London, killing several people and damaging office buildings over a large area. Some surmised this was an attempt to destroy the financial infrastructure of the country, thereby sending shock waves directly to the United States.[28] In 1997, President Clinton, in his zeal to strengthen economic opportunities in Ireland, urged greater American investment in Northern Ireland. But at that time American industry had many second thoughts about placing its people or assets in harm's way, and such investments will surely be slow to come.[29] Now, the warring factions seem committed to settling the tumult once and for all.

As for the rest of Europe, many Americans have an affinity for Paris; they seem somewhat more disturbed by the frequent spates of bombings in the City of Lights. Many potential travelers to Europe surely changed plans during 1995 when terrorist acts accelerated. (See Table 19.) With both Algerian and Corsican terrorists on the loose, the authorities in Paris have taken a number of intrusive counterterrorist measures. Security forces are empowered to operate in airports and train stations, as well as in stores and other public places, to search people more or less at will, and to take other measures justified on the basis of public safety.[30]

TABLE 19.　Four Months of Terrorist Incidents in France in 1995

- July 25: Bomb explodes in regional transit system at St. Michel station in heart of Latin Quarter; eight people killed and more than eighty injured.
- August 17: Bomb laden with nuts and bolts explodes in trash can near Arc de Triomphe; seventeen people injured.
- August 26: Authorities discover bomb planted on high-speed train track north of Lyon; bomb does not detonate.
- September 3: Detonator on a homemade bomb explodes in open market near Place de la Bastille; four injured.
- September 4: Powerful bomb that had failed to explode found in public toilet near outdoor market in southern Paris.
- September 7: Car bomb explodes outside Jewish school in Lyon suburb ten minutes before school lets out; fourteen people injured.
- October 6: Gas canister containing nuts and bolts explodes in a trash can near Maison Blanche subway station in southern Paris; twelve injured.
- October 17: Bomb explodes aboard train on regional subway line between St. Michel and Musee d'Orsay; twenty-nine injured.

Source: Bombings in France, 1995, *The Washington Post*, December 4, 1996, p. 6.

The bombing incidents around the world have mounted into the tens of thousands per year *outside* war torn areas. Some of these bombings are clearly terrorist-inspired; many could not fit even with the broadest definition of terrorism. The characterization of many incidents, whether they be in a church in Germany,[31] scattered throughout Austria by a Eurobomber,[32] or on the buses of Israel, depends on the definition of terrorism, a definition lost upon the victims.

While attention focused overseas, the 1993 bombing of the World Trade Center and the 1995 destruction of the Federal Building in Oklahoma City underscored the vulnerability of buildings throughout the United States. In June 1996, the bombing of the U.S. Air Force housing complex in Saudi Arabia brought home the global nature of the danger to Americans.

Despite greatly accelerated efforts of federal and local authorities to shore up defenses against such acts, repetitions seem inevitable. The challenge is to minimize the number of incidents and to reduce the amount of destruction when an incident occurs—regardless of motive and means of destruction.

Throughout Washington, D.C., the Oklahoma City bombing stirred memories of the notorious book, *Turner Diaries: A Novel*, originally published in 1978. Civil servants in every corner of the city be-

11. Igor Kurchatov, father of the Soviet nuclear program, keeps watch over the atomic city of Snezhinsk—and the author's delegation visiting from the International Science and Technology Center in Moscow, 1994.

gan to wonder whether the sensational depiction of a four thousand-pound truck bomb made from ammonium nitrate and fuel oil exploding in the underground garage of FBI headquarters in Washington was indeed a preview of real-life incidents to come.[33]

Thus, I was not surprised by comments I encountered about government buildings as targets of domestic terrorism when I visited the U.S. Customs Service in late 1996. While I was interested in their recent experiences in intercepting contraband coming into the country, senior officials were preoccupied with the news that their group would be moving into the new Ronald Reagan office building just around the corner. To them it made little sense to move a law enforcement agency into a building with a huge underground parking garage. The vulnerability of such a building had been made clear to them in New York City. Several officials had requested transfers to field offices where more modest office buildings make less inviting targets.

The State Department building facing C Street has long had formidable barriers leading into the underground garage. These have been supplemented with additional security patrols. As a frequent visitor to that building, I exercise great caution when I walk past the entrance to the garage. Every time a vehicle enters or leaves the garage, I am reminded of a fire station with procedures to clear a path from a closed garage onto a main thoroughfare for an emergency vehicle responding to an urgent call for help.

Of course strolling past the White House has become a favorite pastime for tourists who no longer need to compete for space with vehicles now blocked from using this stretch of Pennsylvania Avenue. Manufacturers of the concrete barriers that can be quickly emplaced and moved as necessary—with the help of a crane—have certainly found a market for their truck-resistant products in Washington. Many other cities have rapidly followed suit in building imposing barricades around federal buildings. However, I am not convinced of the total invincibility of our government security services.

Seven years before the Oklahoma City incident, the National Research Council prepared for the federal government recommendations concerning the security of federal buildings. A key recommendation was the need to prevent the parking of vehicles immediately adjacent to such buildings. *Nobody listened.*[34]

According to FBI data, 1,000–2,000 bombing incidents take place in the United States each year. The FBI classifies a few of these as terrorist incidents. Of the more than 2,200 incidents and hoaxes during 1995, for example, only one was classified as international terrorism and nine were considered domestic terrorism.

What were the rest? The largest category—with 1,469 incidents—is classified as "unknown." This was followed by juvenile

incidents (1,014), other (216) acquaintance/neighbor (86), domes-
tic/love triangle (90), gang-related (89), drug/narcotic related
(36), and racial/bias/ethnic group (28). Still, the FBI, in their reports, features "terrorists" when describing those who are able to
use improvised explosive mixtures in combination with sophisticated fusing systems. Singling out terrorists apparently reflects an
FBI belief that while there will be relatively few incidents, they
will be the big ones wherein the perpetrators are most skilled in
both avoiding detection and mounting an attack.

The FBI data notes that during 1996, the five favorite targets for
bombers were: mailboxes/private property (846 incidents), private
residences (136), automobiles (195), academic facilities (100), and
open areas (159). The most popular types of explosives are: black
powder, smokeless powder, pyrodex (36 percent); pyrotechnics/
fireworks (30 percent); and chemical mixtures (17 percent). These
data further suggest that terrorists would rely on high explosives
with greater power such as C-4, TNT, and dynamite that are not reflected in the data on usual types of events.[35]

Pipe bombs tossed by juveniles may well be the most common practice confronting the FBI's Explosives Unit during the
next several years. The increased fear over terrorism may simply
feed the appetites of hoaxers who enjoy challenging authority. It
seems unlikely, for example, that the five-page letter filled with
misspellings and rambling paragraphs directed to "Miss America,
women with blond hair, and CIA spying for the KGB," which
supported a bomb hoax in 1997, was the work of a skilled terrorist.[36] Nor is it likely that the man who blew his hand off while
booby-trapping a New York City apartment with homemade
bombs had ominous international linkages.[37] Still even when the
FBI found a drum containing baking soda loaded onto a fertilizer
truck, the possibility of such a mixture exploding was sufficient to
trigger an FBI terrorism advisory announcement warning of such
concoctions.[38]

In 1997, a wave of letter bombs added to the difficulties in
protecting Americans.[39] Whether delivered by the post office,
courier service, or a sly individual, these often lethal packets take
their toll in killing or maiming innocent recipients. During 1998,
the wide publicity attendant to the trial of Unabomber Theodore
Kaczynski, who had sent his share of letter bombs in years past,

should have reminded Americans to be careful in opening parcels from unknown sources.

Compared with some countries, the United States has lax controls on the purchase of explosives, including detonators and boosters. Many states permit purchases without background checks or even verification of a customer's identity. Even though most high explosives used in bombings are stolen, frequently from unguarded inventories of farmers or small businesses,[40] some of the most dangerous chemicals used in explosives (such as nitrates and chlorates) should be controlled. Controls might include keeping records of sales. Licensing of sellers and permitting of buyers should be considered for some explosives. Inspections of storage areas seem important. Some cases may warrant banning sales altogether.[41]

Such steps would encourage purchasers of explosives to safeguard them more diligently and would thereby help deter bombings. They would increase the difficulty in obtaining explosives and their components, improve the likelihood of detecting criminals during the manufacture or transportation of bombs, and facilitate the tracing of explosives used after a blast.[42]

Bombs will continue to be favorite devices of terrorists in the United States and all over the world. The damage from the small bombs may be minimal, but seemingly minor incidents may eventually become test runs for more serious hits. A particularly worrisome aspect of the small bombings is the lack of vigor of the public outcry over deaths of the innocent victims, a callousness that could unwittingly encourage future terrorists to use greater destructive power for a more devastating effect.

With a much longer history than airplane hijacking, hostage taking continues to serve the purposes of terrorists, with Colombia leading in number of daily kidnappings.

CNN featured hostage seizures in at least three areas of the world during late 1996 and early 1997. First, in December 1996, fourteen members of the waning but still active Tupac Amaro Revolutionary Movement (MRTA) held 500 diplomats and government officials hostage in the residence of the Japanese ambassador in Lima, Peru, for up to four months. Fortunately only one hostage was killed and that was during the raid that ended the seige.[43] Then, in March 1997, four Russian journalists were held captive for three months in Chechnya, and released in June as part of the reconciliation process between Russia and dissident Chechen leaders.[44] Shortly thereafter, a three-member Russian TV crew disappeared in the same region with their release eventually secured in August 1997 by a ransom of $1 million.[45] Finally in April 1997 in Texas, a militant group proclaimed the sanctity of property belonging to the newly proclaimed Republic of Texas. They took hostages, who became the centerpiece of a standoff between separatists and police.[46]

To most terrorism experts, the drama of the hostage rescue in Lima unfolded as a brilliant success story based on decades of trial and error responses to hostage-taking. The strategy of Peruvian President Alberto Fujimori followed a dual track: the public operation and the clandestine one. (See Table 20.)

From the outset, Fujimori saw little chance of resolving the standoff peacefully. But he gave negotiations a try, if only to mask preparations for the assault. He arranged also safe passage to Cuba for the rebels, who refused them.

Fujimori was determined to take full responsibility—and credit—for resolving the situation. Thus, he resisted many offers of foreign assistance in the rescue operation. He did accept some technical help, using CIA-provided matchstick-size two-way microphones that allowed intelligence officials to communicate with military commanders being held inside the residence. These devices were concealed in books, guitars, and thermos bottles sent into the residence, purportedly at the requests of the families.

The raid has been described as one of the most successful hostage rescue missions of modern times. It occurred as the guerrillas were playing indoor soccer. Just as they bellowed "goal," the floor exploded under their feet and five of them were instantly killed. At the same time, other charges blasted more openings from the freshly dug tunnels into the interior of the residence and still others blew holes in each side of the building's exterior. Commandos poured out of the tunnels, firing as they ran.

As the smoke cleared inside, the commandos organized a parade of hostages on their hands and knees, like a trail of ants, as one hostage put it. They crept onto a balcony and down an outside stairway to safety.

TABLE 20.   Hostage Crisis at the Japanese Embassy in Lima, Peru

Public Events
- December 17: Fourteen guerrillas seized five hundred hostages at the residence, with the rebels threatening to kill hostages unless the government released jailed comrades. The deadline passed as the government stalled.
- December 19: The International Red Cross was designated as intermediary, and the rebels began the process of releasing batches of hostages.
- December 24: Uruguay's ambassador was released after his country freed two rebels held in Uruguay. Peru lodged a diplomatic protest in Montevideo.
- January 15: The rebels agreed to talks provided freedom for jailed comrades was on the table to be discussed.
- February 1: Fujimori agreed to talks but insisted that jailed rebels would not be freed.
- February 11: Talks began.
- March 3: Fujimori traveled to Cuba to secure Castro's offer of asylum if needed to end the standoff.
- March 6: Rebels broke off talks, claiming a secret tunnel was being constructed into the residence which the government denied.
- March 13: Indirect talks through mediators broke down over rebel demands for release of hundreds of jailed guerrillas.
- April 22: Peruvian forces stormed the residence, and rescued seventy-one hostages with one hostage, two commandos, and all fourteen rebels killed.

Secret Events
- December 17: Fujimori immediately ordered military commanders to begin training for a rescue mission.
- Late December: A full-scale mock-up of the residence was built at a naval base, and intelligence overflights of the residence by U.S.-provided surveillance aircraft began.
- January: Construction of 190 feet of tunnels began.
- Mid-March: A tiny transmitter was smuggled into the residence, and several hostages became part of the military rescue conspiracy.
- Late March: Intelligence operatives learned that the rebels were growing slack in security and that they played soccer every afternoon.
- April: As tunnels neared completion, tiny periscopes with listening devices were inserted into the floors of the building.
- April 21: Commandos were deployed in five tunnels.
- April 22: The commandos, knowing precisely where each hostage and guerrilla was, set off plastic explosives blowing holes in the walls and floors and then rescued the hostages within sixteen minutes.

Source: Peru Crisis: Public and Secret Missions, *The Washington Post*, April 27, 1997, p. 28.

Soon all fourteen terrorists were dead. One Peruvian officer, who visited the site, estimated each terrorist had at least five hundred bullets in him. While this was probably an exaggeration, the rescuers clearly were under instructions to shoot on sight, and the shots were numerous. Whether some of the rebels had tried to surrender and were summarily executed will remain an unresolved controversy. What is certain, the "no prisoners" strategy has sent a strong message to other terrorists within and outside the country.[47]

The success of this rescue operation differed dramatically from the botched attempt of the United States Air Force and Army to rescue fifty-two hostages held at the American Embassy in Teheran for 444 days. The Ayatollah Khomeini and his Islamic fundamentalists took the hostages November 4, 1979.[48] Whether such a mission should have even been attempted will be debated by historians for decades. From my perspective, the execution of the operation certainly was not in the hands of skilled professionals.

The rescue plan called for launching a helicopter assault from a remote desert staging area within striking distance of Teheran. Training was carried out at Area 51, the super-secret air force complex to the north of Las Vegas, not far from my home at the time. However, the only people who thought that Area 51 was indeed secret were officials in Washington. Passersby watched unmarked aircraft shuttle work crews from the Las Vegas airport to the desert hideaway every day and the local press carried frequent stories, although probably not always accurate, about the goings-on at the site. Consequently, when the details of the bungled operation came to light, Nevadans felt some sense of attachment to the effort.

A fatal technical flaw in the mission was attributed to the helicopters stirring up loose sand which clogged their rotary mechanisms, resulting in the grounding of three of the helicopters sent to the scene. The inappropriateness of Area 51 as a training base became clear too late. The Nevada desert is rock hard. Even the most violent whirling of helicopter blades results in only a thin layer of loose sands being stirred into the atmosphere. On the other hand, the Middle East deserts are soft; large volumes of sand can be stirred by helicopter blades.

Rescue missions will undoubtedly be at the top of the response options whenever a serious hostage situation emerges. Governments will immediately focus on rescue by negotiation, stealth, or force. As ambassadors and embassies continue to be high-profile targets for hostage situations, the art and practice of hostage rescue will become no less complicated than the art and practice of traditional diplomacy. Both require horse trading and face saving.

As reported by an American diplomat who for five days was a hostage at the Japanese Ambassador's residence in Peru, refusing to negotiate unacceptable demands with terrorists was definitely the way to go. He noted that President Clinton's message for other governments to stop pressuring Fujimori to swap hostages for prisoners was broadcast on TV sets throughout the residence. Immediately, the captors recognized the devaluation of the hostages as prizes in a high stakes game.[49]

Turning to the increasing threat of individual terrorists, the definition of assassination has become somewhat muddied. To most of us, the word implies the planned murder of a specific individual; whereas terrorists often kill whomever is in their line of fire. The well-known shootout in 1993 at the entrance to the CIA offers the latter perspective in describing terrorist assassinations. In this case, the CIA as an institution represented the general target of the assassin, while the actual victims were randomly picked because they were in the line of fire. Regardless of the narrowness or scope of the definition, the result was the same: two dead bodies.

When I moved to Washington, D.C. in the 1950s, no one would dare mention in public which building housed the Central Intelligence Agency. In the 1960s, my friends outside government surmised that the agency had moved across the Potomac River into Virginia. In the 1970s, all knew that a certain cutoff from the George Washington Memorial Parkway would lead not only to the advertised Federal Highway Research Center but also to CIA headquarters. By the 1980s, the turnoff from Dolly Madison Boulevard in Langley into the agency had gained the stature of a national historic landmark with requisite signage.

On January 25, 1993, television stations across the country reported an incident at that very intersection. A man wielding an assault rifle emerged from a car, walked between the cars waiting in the left lane to turn into the agency, and fired seventy bullets at the vehicles. There was little doubt that this was an assault on CIA employees. Indeed, the assailant succeeded in killing two CIA employees while wounding two other employees and injuring one other person.[50]

Mir Aimai Kansi (now known as Kasi), a Pakistani national living in nearby Reston, was soon identified as the primary suspect. Prior to the attack, on four separate occasions he had purchased a variety of weapons at a gun shop. He left for Pakistan the day following the murders, carrying no luggage. His roommate reported to the police that shortly before the shooting, Kansi had complained about the treatment of Muslims in Bosnia and wanted to make a statement by shooting up the CIA, the White House, or the Israeli Embassy.

Within two weeks, Kansi had been charged with the attack after an AK-47 rifle and other evidence were found in his apartment. One week later, the FBI placed Kansi on its Ten Most Wanted list. At the end of March the FBI informed the State Department that the attack was "an act of international terrorism." After initial hesitation, the State Department announced the reward for information leading to the arrest of Kansi had been increased from $100 thousand to $2 million.

For four years, U.S. agencies carried out a search for Kansi, working with a sometimes reluctant Pakistani government. Unofficial reports suggest he initially fled to his native Baluchistan, a loosely governed region of Pakistan near the Afghan and Iranian borders. In 1993, the Pakistani army raided the Kansi family compound after FBI and CIA teams heard that the fugitive was in the area. But he was not there and was believed to have escaped into Afghanistan.

Apparently, over time, local Afghans or Pakistanis with no family ties to Kansi became interested in the $2 million reward and provided useful information on his whereabouts. Also, Pakistani intelligence sources were reported to be helpful as were other informants assisting the U.S. government in counter-narcotics activities in the region.

The endgame for Kansi is believed to have begun when Afghan collaborators offered to assist in his capture and lured him to a seedy hotel near the Pakistani city of Quetta, where his family owns real estate and businesses. During the night, four FBI agents burst into Kansi's room, slammed him to the floor, and took him into custody. He was hustled into a waiting airplane and taken to an undisclosed location. Two days later a U.S. Air Force plane flew Kansi and his captors to the United States. In early 1998, he was sentenced to death in a Fairfax, Virginia, courtroom.[51]

In a more prolonged affair, the Canadian town of Charlottetown (population 30 thousand) and its environs were rocked for a decade by pipe bombs. Against a mosaic of pronouncements advocating individual freedom, condemnations of government bodies, and displays of swastikas, a former high school chemistry teacher, Roger Bell, delivered the bombs. He bombed the courthouse in 1988, a park in 1994, the regional legislature in 1995, and a gas depot in 1996. He used ordinary gunpowder packed into plumbing pipes and detonated with digital timers.

When arrested, he offered little as to his motivations. The intentions of this loner remain unknown. He did not target individuals as had the Unabomber but rather focused on garnering public attention. In that, he succeeded. According to locals, he had simply become a totally different man after resigning from his teaching post: he went into seclusion, spent all his money, and started setting off bombs.[52] Let us hope he was simply in the category of a highly intelligent crazy with few clones on the horizon.

I have been discussing types of terrorism that have dominated the international landscape for the last forty years. Aircraft banditry, buildings destruction, hostage seizing, and cold-blooded murders will continue to be tools of the trade. New brands of suitcase plastic bombs, poison venom assassination canes, and incendiary pocket diaries will become so intricate that even James Bond and his mentor "Q" would be impressed.[53]

Law enforcement agencies have learned a great deal as to how to limit the damage and pursue the perpetrators; no doubt they will continue to keep their guards high. They will not win all of the battles, but to date the successes of terrorists in using traditional approaches have not changed political institutions in a significant fashion.

However, still more ominous dangers accompany the expansion of the physical infrastructures supporting economic and related activities in the United States. As they grow, they become more interdependent, and they also become more vulnerable.

Gas pipelines, electric power grids, aircraft traffic control systems, and emergency response networks are among the terrorist-prone targets with multiplier effects. In 1996, a presidential commission was established to recommend steps to improve the safeguarding of these and other systems. There has always been concern over the physical threats to tangible property associated with the networks. With the increased reliance on electronic circuitry to control the operations of so many public and private enterprises, the vulnerability of the systems to electronic warfare has moved to the top of the counter-terrorism agenda. (See Table 21.)[54]

The commission has advocated a shared response approach since the private sector operates over 90 percent of the networks. Public and private sector owners and operators of these systems should upgrade security, while appropriate government agencies should collect and disseminate to owners and operators information about terrorist groups and the possible tools of disruption and destruction they might be using. Also, the agencies should expand research on the effectiveness of security systems. They should also take steps to update criminal, privacy, and international laws that have not kept pace with advancing technologies.[55]

Of special concern in terms of protecting lives are transportation hubs and other facilities where people congregate. A foiled

TABLE 21.   Our Nation's Critical Infrastructure

- *Telecommunications*: systems that support transmission of electronic communications.
- *Electric Power Systems*: generation stations, distribution systems, and transportation and storage of fuel.
- *Gas and Oil Production, Storage, and Transportation*: holding, refining, and processing facilities and distribution systems.
- *Banking and Finance*: retail and commercial organizations, investment institutions, exchange boards, trading houses, and reserve systems.
- *Transportation*: aviation, rail, highway and aquatic vehicles, conduits, and support systems.
- *Water Supply*: reservoirs, aqueducts, filtration systems, pipelines, cooling systems, and systems for dealing with waste waters and fire fighting.
- *Emergency Services*: medical, police, fire, and rescue services.
- *Continuity of Government Services*: federal, state, and local agencies including support of public health, safety, and welfare.

Source: President's Commission on Critical Infrastructure Protection, 1997

plan to cause havoc in the New York City subway system in the summer of 1997 underscored the danger. Fortunately, a brave, but shaken informant using broken English and sign language, convinced New York police officers that his roommates were about to cause carnage that would rival the massacre at Jerusalem's largest market a few days earlier. In a bloody predawn storming of a Brooklyn apartment, the authorities seized the conspirators who were plotting subway blasts from exploding pipes they had packed with nails, bullets, and gunpowder.[56] What would have happened had that roommate not gone to the authorities? And how quickly most of us have already forgotten about this near miss, a simple plan that a copycat perpetrator could replicate in any one of a number of cities of the United States.

Yet the time bombs of greatest concern may take shape in computer software. These would attack the electronic networks supporting the many institutions discussed above. As precursors of more to come, the PLO planted a virus in a computer system at Hebrew University; Japanese terrorists have attacked computer systems that control commuter trains; and the Red Brigade's manifesto calls for destruction of computer systems. The list goes on. I have already noted many other dimensions of emerging cyberterrorism, and other books devoted exclusively to this topic are beginning to appear.

No matter what century, terrorists seem to be first on the block with their kind of new technology—be it a shiny dagger, innovative detonators that trigger explosives at predetermined times, or software to penetrate protected computer files. Odds are that the armories of the criminal conglomerates of the future will stock all of the traditional tools and many new ones that promise greater destruction. With such arsenals, increased access to knowledge and money will be the key determinants of the level of danger posed by terrorists. Whether acting alone, in groups, or as state-sponsored mercenaries, they need expertise both to use their weaponry and to outsmart their opponents. They also need money, lots of it, to implement their plots. Perhaps the good news for those who must fight terrorism is that the money trail, the brain trail, and the equipment trail all may provide clues to clandestine activity incubating in evil hatcheries.

In the meantime, the response of the U.S. government to the threat of terrorism has been extensive. Though we can never fend off all threats, we must reduce them to tolerable levels. If the money and knowledge of people attracted to violent crime could be redirected to productive pursuits, that would be awesome indeed. Such redirection is the challenge of the next century. I will later suggest some long-range ideas, but it is imperative that at the same time we punch the fast-forward button to expand efforts we already know to be effective.

CHAPTER 9

The Growing Arsenal to Combat Terrorism

Terrorism is the enemy of our generation,
and we must prevail.
President William Clinton, 1996

Let's have a Fourth Amendment [search and seizure]
that works in the information age.
FBI Director Louis Freeh, 1997

The ideal detection equipment for local bomb squads and
first responders should cost no more than $10 and be so
small it can be worn like a badge or on a belt or stuffed in
a pocket. It should be so reliable it never misses the real
thing and only [has] false alarms once or twice a year.
Office of Science and Technology,
National Institute of Justice, 1997.

The steps we are taking today of cracking down on fundraising for terror and of banning terrorists from our shores are steps we urge other countries to take within their jurisdictions. By steadily reducing the habitat in which terrorism thrives, we can hope to make terrorists first an endangered species, and ultimately, an extinct one.[1]

These were the ringing words of Secretary of State Madeleine Albright at a press briefing on October 8, 1997. Under a provision of the *Anti-Terrorism and Effective Death Penalty Act of 1996*, she had designated thirty groups as foreign terrorist organizations. (See Table 22.) It is now a crime to provide funds, weapons, or other types of tangible support to any of these designated organizations. Members and representatives of the organizations are ineligible for visas to enter the United States. Any funds the organizations have in the United States will be frozen.[2]

The secretary's comments added teeth to the 1996 Act that identified a number of initiatives to clamp down on terrorism. Her promise of swift and deliberate action signaled that the act would be much more than just a wish list to appease those calling for government action.

TABLE 22.   Foreign Terrorist Organizations

- Abu Nidal Organization (ANO)
- Abu Sayyaf Group (ASG)
- Armed Islamic Group (GIA)
- Aum Shinrikyo (Aum)
- Euzkadi Ta Askatasuna (ETA)
- Democratic Front for the Liberation of Palestine-Hawatmeh Faction (DFLP)
- HAMAS (Islamic Resistance Movement)
- Harakat ul-Ansar (HUA)
- Hizballah (Party of God)
- Gama'a al-Islamiyya (Islamic Group, IG)
- Japanese Red Army (JRA)
- al-Jihad
- Kach
- Kahane Chai
- Khmer Rouge
- Kurdistan Workers' Party (PKK)
- Liberation Tigers of Tamil Eelam (LTTE)
- Manuel Rodriguez Patriotic Front Dissidents (FPMR/D)
- Mujahedin-e Khalq Organization (MEK, MKO)
- National Liberation Army (ELN)
- Palestine Islamic Jihad-Shaqaqi Faction (PIJ)
- Palestine Liberation Front-Abu Abbas Faction (PLF)
- Popular Front for the Liberation of Palestine (PFLP)
- Popular Front for the Liberation of Palestine-General Command (PFLP-GC)
- Revolutionary Armed Forces of Colombia (FARC)
- Revolutionary Organization 17 November (17 November)
- Revolutionary People's Liberation Party/Front (DHKP/C)
- Revolutionary People's Struggle (ELA)
- Shining Path (Sendero Luminoso, SL)
- Tupac Amaru Revolutionary Movement (MRTA)

Source: Press release of the Office of the Coordinator for Counterterrorism, U.S. Department of State, October 8, 1997.

Other directives in this legislation are:

- Deportation procedures for suspected terrorists are to be simplified.[3]
- Plastic explosives are to be marked for easier detection.
- U.S. foreign assistance cannot be provided to countries that aid or provide lethal equipment to terrorist states.
- U.S. citizens can sue terrorist states for damages arising from terrorist-related activities.
- Exports of defense-related items are to be denied to nations not cooperating fully with U.S. counterterrorism efforts.
- Foreign airlines using U.S. airports must adopt the same security standards as U.S. airlines.
- Broader categories for classifying items as nuclear or biological agents are proscribed.
- U.S. extraterritorial jurisdiction over terrorist acts committed in the United States and abroad is expanded.[4]

The U.S. Congress and the executive branch are taking an aggressive stance against terrorism, funneling new resources and enlisting highly skilled personnel for the battle. For example, the President's 1996 $1 billion counter-terrorism initiative included funds, subsequently appropriated by Congress, for dozens of programs from airport security to physical protection of American soldiers to expanded capabilities of the courts to process terrorism cases. Additional programs of more than fifty government agencies are included in other appropriations packages embraced by Congress. These programs warn terrorists that the United States means business. (See Table 23.)

Still, the administration has been disappointed that Congress has not granted more governmental authority to track down terrorists before they strike. For example, Congress rejected proposals to expand federal agency authorities for emergency wiretaps in terrorist cases, for tracing phone calls, and for increased access to hotel and motel records, all pursuant to court authorizations.[5] Such authorities are available under legislation to combat drug trafficking and organized crime. Apparently, our legislators simply did not want to push the civil liberties envelope any further,

TABLE 23. New Counterterrorism Measures Adopted by the Clinton
Administration

- Upgraded airport security through new devices for screening carry-on and
  checked baggage, new technologies for inspecting international air cargoes, additional canine teams, better passenger profiling, and expanded security forces.
- Improved bomb detection through studies of the feasibility of tagging and licensing explosives, increased inspections of explosive manufacturing facilities,
  expanded training for explosive detection specialists, and assessments of previously encountered devices.
- Increased staff for the FBI to assess vulnerabilities in the physical infrastructure
  of the country, to improve daytime and nighttime overhead surveillance of
  suspicious activities (a new Project Nightstalker), and to expand technical capabilities to address nuclear, chemical, and biological threats.
- Better physical protection overseas for American troops in the Persian Gulf region, for senior diplomats, and for diplomatic and trade offices.
- Expanded capabilities of U.S. attorneys and courts to handle additional work-
  loads generated by counterterrorism measures.
- Reinforcement of many U.S. federal buildings, particularly those occupied by
  law enforcement agencies.
- Contingency funds to respond to unanticipated events.
- Expanded efforts to detect illegal exports of relevance to weapons of mass destruction, to prevent nuclear smuggling, and to respond to nuclear incidents.
- Expanded efforts and improved coordination among intelligence collection
  agencies.

Source: "White House Fact Sheet on Counter-terrorism Measures," available from U.S. Information Agency, May 19, 1997.

or perhaps they harbor new levels of mistrust of the behavior of
federal agencies.

Nevertheless, the recent responses of the U.S. government
to the emerging threats of international terrorism can genuinely
be described as *awesome*. Small armies of talented specialists
throughout Washington are devoting both longstanding and
newly discovered careers to countering these national security
threats, with $7 billion of federal government funds specifically
earmarked for antiterrorism programs each year. There is an even
more expensive track of activity to battle narcotrafficking, again
involving several dozen agencies, with federal funds exceeding
$10 billion annually to interdict the supply routes.[6] If the tens of
billions of dollars devoted by the Departments of Defense, Energy, and State to counter proliferation of weapons of mass destruction are added, the total annual expenditures related to
containing international terrorism exceed $50 billion and are

growing. This doesn't include the funds funneled through bilateral and multilateral foreign assistance programs; many of these programs concentrate on helping poor countries reduce motivations for terrorism. In particular, programs supporting moves toward more democratic governance and improved economic opportunities can begin to offer alternatives to the population.

This explosion of government initiatives collectively respond to perceptions and predictions of likely developments around the globe. In late 1997, the Department of State released the government's assumptions concerning relevant trends during the next five years. These assumptions are generally consistent with the forecasts that I have presented, although in some respects the government seems less pessimistic as to the dangers facing the United States. The difference may rest, at least in part, in my longer time horizon and also in the department's greater reluctance to articulate undesirable developments, official speculation that could have considerable political and economic repercussions in other countries. (See Table 24.)

Overseas, American Embassies house increasing numbers of officials from our law enforcement agencies. Our envoys engage in nonstop negotiations on terrorism issues at the United Nations, within other international organizations, and at regional meetings. American ambassadors are continuously consulting foreign counterparts in dozens of capitals in countries that are targets or sources of terrorist plots. The transoceanic data links among the international terropolice flow with information about the movements of dangerous people, diversions of money, and the smuggling of contraband. Still, we are a long way from having a good handle on many terrorist groups and are often surprised when their clandestine schemes suddenly erupt into transparent reality.

On the domestic front, coordination among federal and local officials to deter and respond to perpetrators in our own homeland has become a new way of life in Washington. Congressional hearings fill meeting rooms of the Dirksen and Rayburn buildings every week. Fire chiefs, bomb disposal experts, and health department officials from throughout the country block out more and more time in their daily planners to participate on a multitude of Washington-based task forces and advisory committees, attend

TABLE 24.   Strategic Assumptions of the Department of State (1997–2002)

International Terrorism
- U.S. officials, facilities, and citizens will remain high-priority targets of terrorists.
- International cooperation to combat terrorism will continue to increase.
- State support for terrorism will continue to decline, but the number of unaffiliated, ad hoc, and new terrorists will increase. Religiously motivated and sectarian terrorism will grow.
- The number of international terrorist incidents will continue to decline, but terrorists will seek to increase casualties and damage by using more lethal explosives.
- The danger that terrorists will employ chemical, biological, or nuclear material will grow.
- Terrorists will attack less-protected targets, including vulnerable communications systems and infrastructures.

Weapons of Mass Destruction and Destabilizing Conventional Arms
- When the United States cannot fully deny to hostile states the technology for weapons of mass destruction, it can retard the rate at which advanced technologies appear in their arsenals or deter their use.
- Uncertainty about the stability of Russia's military will persist, but Russia will continue to fulfill arms control and nonproliferation commitments.
- Arms control treaties have become increasingly complex and their requirements more intrusive. Negotiations and implementation require high expertise and long-term commitments to implementation and compliance.

International Crime
- Transnational crime is undergoing a significant evolution, particularly with the appearance on the world scene of criminal organizations from Russia, Asia, and elsewhere.
- Increasingly, foreign criminals will seek opportunities in the United States, and American criminals will seek opportunities abroad.
- Improving the criminal justice systems of foreign governments will contribute to their ability to control their own crime problems and to work with the U.S. on international crime issues.

Drug Trafficking
- In order to reduce the entry of illegal drugs into the United States, it is necessary to reduce foreign production.
- The supply of illegal narcotics from abroad responds to demand from the United States.
- Although foreign governments have an interest in counternarcotics cooperation with the United States, limited institutional capacity, along with social, political, and economic factors—including corruption—will remain major constraints.
- The Western hemisphere will remain the major foreign source of illegal drugs with the Mexican–U.S. border and Puerto Rico the major entry points.

Source: *United States Strategic Plan for International Affairs*, U.S. Department of State, September 1997.

training courses, and speak at town meetings. Terrorism is a hotbed of action for law enforcement agencies and responders at all levels.

Government officials repeatedly emphasize the U.S. terrorism policy of no concessions to terrorists, formulated largely in response to hostage situations: no ransom payments, no releases of prisoners, and no changes in policies in response to demands. American officials constantly press other governments to follow suit. Our diplomats stress the importance of unrelenting pressure on state sponsors of terrorism and the need to use all available legal mechanisms to deter and punish international terrorists. They underscore the significance of programs of assistance to third world governments—to improve their capabilities for combating violence through their legal systems and through their police and their armies.[7]

Is it really possible to hang tight when terrorists are in your face with guns, bombs, and threats of massive killings, to refuse to comply with demands when many lives as well as critical issues are at stake? In principle, we should not cave in. But in practice, there have been U.S. deviations when lives were at stake. On more than one occasion, in dealing with events in the Middle East when Americans were held captive during the 1980s, we bent to the demands of captors to implore Israel to change its policies toward punishing Muslim terrorists and seizing disputed land.[8] Before applying military or economic pressure on suspected state-sponsors of terrorism to change course, we consider many interests in addition to counterterrorism. They include investments of American business, interests of ethnic groups in the United States with special ties to the countries of concern, and views of allies as well as adversaries.

Despite occasional inconsistencies, the official positions and practices of the United States are pacesetters for the world. The U.S. government leads the pack in terms of proactively confronting violent adversaries at home and abroad. At the same time, interventions into the affairs of governments, groups, and individuals must be harnessed lest they run roughshod over the principles of international law and human rights.

However, the U.S. government insists on a terrorism policy distinct from those that apply to weapons proliferation, organized crime, narcotrafficking, and money laundering. At times, the delineation sets up confusing boundaries that diminish the collaboration so essential in disrupting interrelated criminal activities. When it comes to implementing programs, those in charge should be encouraged to lower barriers between agencies and pursue the cooperative strategies most likely to succeed. Indeed, fragmentation of policy and program responsibilities led the former director of the CIA to offer a derisive, "Ha,ha," when critiquing U.S. organizational readiness to respond to the terrorist danger. We will return to the importance of a more contemporary conception of terrorism later.[9]

Throughout our earlier discussions, we identified many actions taken at the international, national, and local levels to respond to specific types of threats to Americans and American institutions. In addition, despite fractionalized programs, a number of important cross-cutting measures are forming a cohesive defensive lattice woven tightly enough together to choke off many threats of violence.

The Congressional Research Service has assembled a two-inch thick tome of treaties and agreements to tackle international terrorism.[10] Eleven agreements have been formally classified as "terrorism conventions"; they are directed to hijackings, bombings, hostage situations, and other acts of greatest concern in decades past. (See Table 25.) However, they are but a part, less than half, of the larger legal umbrella to thwart terrorists. The other agreements range from controlling exports of sensitive materials to reducing money laundering to foreswearing the use of chemical and biological weapons. Of course, many of the nations of greatest concern have not signed some of the agreements; those that have, can erect formidable barriers for rogue rulers who do not want to join the mainstream of civilized society.

Since the mid-1990s, terrorism has moved to the top of the agenda of summit diplomacy. Following the assassination of Israeli Prime Minister Rabin, President Clinton chaired a meeting in Egypt in mid-1996 of twenty-nine national and world leaders. His purpose was to stimulate stronger counterterrorism initiatives and thereby contribute to the Middle East peace process. The attendance and agenda were impressive although the long-term im-

TABLE 25.   Status of Eleven Terrorism Conventions

- *Convention on Offenses and Certain Other Acts Committed on Board Aircraft*—developed in 1963, entered into force December 4, 1969.
- *Convention for the Suppression of Unlawful Act Seizure of Aircraft* (Hijacking Convention)—developed in 1970, entered into force October 14, 1971.
- *Convention for the Suppression of Unlawful Acts Against the Safety of Civil Aviation* (Sabotage Convention)—developed in 1971, entered into force January 26, 1973.
- *Protocol for the Suppression of Unlawful Acts of Violence at Airports Serving International Civil Aviation*—developed in 1988, entered into force August 6, 1989 (for the United States: November 18, 1994).
- *Convention for the Suppression of Unlawful Acts Against the Safety of Maritime Navigation*—developed in 1988, entered into force March 1, 1992 (for the United States: March 6, 1995).
- *Protocol for the Suppression of Unlawful Acts Against the Safety of Fixed Platforms Located on the Continental Shelf*—entered into force March 1, 1992 (for the United States: March 6, 1995)
- *Convention on the Marking of Plastic Explosives for the Purpose of Detection* (Plastic Explosive Convention)—developed in 1991, not yet in force, awaiting one more signator confirming ratification by that government.
- *Convention on the Prevention and Punishment of Crimes Against Internationally Protected Persons*—entered into force February 20, 1977.
- *International Convention Against the Taking of Hostages*—entered into force June 3, 1983 (for the United States: January 6, 1985).
- *Convention on the Physical Protection of Nuclear Material*—entered into force February 8, 1987.
- *Convention for Suppression of Terrorist Bombings*—opened for signature January 12, 1998.

Source: Department of State, February 1998.

pact of the meeting remains mired in stop-start mediations with no resolution in sight.[11]

Later that year in Lyons, France, the heads of state of the United States, Canada, Japan, Germany, France, Italy, and the United Kingdom committed to give "absolute" priority to the fight against terrorism. They instructed their ministers to develop practical measures they could implement without delay and could also recommend to leaders of other governments.

Shortly thereafter in Paris, the ministers reaffirmed the commitment to condemn all forms of terrorism, make no concessions to terrorists, and fight terrorism in ways that are consistent with fundamental freedoms and the rule of law. Against this background of diplomatic promises, they set forth many recommendations to expand the portfolio of international treaties for countering political violence and for improving cooperation and information exchange. Prosecution and punishment of perpetrators were high on their list. The ministers underscored the importance of preventing terrorists from obtaining false travel documents—a daunting task indeed—and from finding sanctuaries under the guise of political-refugee status. In addition, while they did focus on preventing terrorist groups from engaging in fundraising activities, *they did not adequately address the important issue of laundering the money that keeps some terrorist groups in business.* As we have seen, Washington has quickly moved forward to stop practices in all of the areas identified at the meeting.[12] (See Table 26.)

In June 1997 in Denver, the seven heads of state were joined by Boris Yeltsin who increased their ranks to the Group of Eight, also known as the G-8, or the P-8, and now simply the Eight. There the ministers delivered their findings in the shadows of a spate of

TABLE 26.   Foreign Ministers' Progress Report at Denver Summit of the Eight

- Preparation of a UN Convention on Suppression of Terrorist Bombing, with special attention to bombings in government buildings, public places, and public transportation.
- Cooperation among states in investigating terrorist crimes involving motor vehicles, including standardized approaches to vehicle identification numbers.
- Improved consistency of airport security standards, including improved explosive detection capabilities.
- Prevention of abuse by terrorists of legitimate rights of political asylum that are set forth in international law.
- Countering the use by terrorists of encryption techniques by allowing authorized government access to coded messages in order to investigate terrorist activities.
- International sharing of U.S. forensic databases.
- Fulfillment by states of their international obligations to prevent terrorists from gaining access to biological weapons.
- Exchange of information on laws to prevent terrorist fundraising.
- Adoption of strong domestic laws to control manufacture, trading, and transport of explosives.
- Prevention of attacks on computer networks.
- Expanded research on technologies to counter the use of high explosives and materials of mass destruction.
- Increased attention to security at major international events.
- Heightened vigilance against terrorism on ships.

Source: Department of State, July 1997.

terrorist bombings in France, Russia, the United Kingdom, the Middle East, and South Asia; they also reacted to the seizure of hostages at the Japanese ambassador's residence in Peru six months earlier. The Eight repeated the well-known litany of granting no concessions to terrorists and refusing to capitulate to the demands of hostage takers. They called for still more terrorism treaties while enlisting more signatories to existing treaties. Their report provides an excellent checklist of those items currently at the top of the international counterterrorism agenda.[13]

While international consultations on terrorism take place in many forums throughout the year, this recent interest of the leaders of the Eight is significant. The ripple effects of their combined efforts at the highest political level should prove to make life much more difficult for terrorists the world over. Their political decisions will not stop acts of violence. They can trigger concerted international efforts in the areas of intelligence, extradition, and public censure, that underscore the resolve of the most powerful nations in the world to squeeze terrorism from all sides.

High on the list of expanded U.S. government efforts to protect Americans are programs to contain the international spread of weapons and materials of mass destruction and to control them wherever they exist, a process of intense international action that began more than fifty years ago. For many years international concerns were directed to preventing heightened levels of all-out warfare. Only in the 1990s did the focus of nonproliferation efforts widen to seriously challenge terrorist use of high-tech byproducts of military preparations for war.

The U.S. Department of Defense believes that "the likelihood of a state sponsor providing a chemical, biological, or radiological weapon to a terrorist group is low." The Pentagon adds, however, that extremist groups with no ties to a particular state might acquire and attempt to use such weapons, as was the case with Aum Shinrikyo. If an attack is launched by an independent group, a response in kind seems out of the question. Retaliation using other means, including conventional weaponry as necessary, would be the order of the day.[14]

As discussed in earlier chapters, the Nuclear Nonproliferation Treaty, the International Atomic Energy Agency, and other international arrangements are in place both to restrict the diffusion of nuclear-related equipment and materials and at the same time to promote cooperation in the peaceful uses of atomic energy.[15] Also, we have seen that international agreements have been reached to limit the spread of biological and chemical agents and equipment to forestall the possibility of terrorist use.[16]

Two other tracks of intergovernmental activity have been directed to strengthening the control of exports of (1) missile technologies and (2) dual-use technologies that could support advanced conventional weapons. The diversion to military programs of dual-use technologies ostensibly acquired for legitimate civilian applications, whether they be rockets for space exploration or lasers for industrial applications, is a central concern. Iran, Iraq, North Korea, and Libya, in particular, have been singled out by the U.S. and other governments as seeking access to sensitive dual-use items as well as military hardware.[17]

These international approaches to limit exports are very significant in deterring the transfer of dangerous items to hostile countries. At the same time, none of the procedures are foolproof. Not all relevant states participate in the agreements; inspection procedures for confirming compliance with international commitments are far from adequate; and appropriate penalties for noncompliance are lacking. The array of agreements that have resulted from long and tedious diplomatic efforts represent key building blocks in fortifying the firewalls that protect America.

Recognizing the remaining vulnerabilities in the architecture of international agreements, the U.S. government is working hard to seal the cracks. For example, in response to warnings of black market interest in uranium and plutonium caches, international meeting agendas include debates over setting standards to specify appropriate levels of physical protection of these materials at individual facilities. These discussions have expanded to include procedures for combatting trafficking in nuclear wastes. Diplomats continue their slow progress in developing an international inspection protocol to ferret out research facilities that could support illicit biological weapons activities. U.S. officials continuously implore their foreign counterparts to ensure that export control decisions of individual states give adequate weight to proliferation dangers even though economic interests may be pressing for prompt exports of sensitive items.

Of course, international agreements alone cannot adequately curtail the access of terrorist groups within individual countries to laboratories and supplies of materials sufficient for mounting high-tech attacks. Indeed, the United States sports all the ingredients terrorists could ever dream of to blitzkrieg the population with chemical or biological agents. The FBI leads the agencies that recognize the country's vulnerabilities, not only to misguided elements of our own population but to the international dimension of foreign terrorists intent on amassing made-in-America weaponry.

With the Oklahoma City and World Trade Center bombings spotlighting locally based assaults, Congress required the president to submit in early 1997 the report *Response to Threats of Terrorist Use of Weapons of Mass Destruction.*[18] This report emphasized the heightened chances of a significant incident in the United States. The increased risk is a direct result of terrorist awareness of the availability of chemical and biological agents. The portability of small amounts of such agents is especially useful for clandestine purposes, as is the potential for a large-scale catastrophe when local responders can neither quickly identify nor contain the effects of such agents.

The report states that the first responders to an attack, usually the fire and police departments, would serve as critical gatekeepers of public safety. Their successful and timely training is now central to the president's agenda, with special attention to safety protection of the responders themselves and a high priority placed on improving the equipment available for detection and analysis of hazardous agents. Local agencies would like to expand inventories of antidotes and other medical supplies, with the Veterans Administration's hospitals being considered as stockpile locations. Also, both federal and local agencies recognize that field exercises to improve response capabilities and coordination mechanisms must be expanded.

In other areas, the report pointed out the need to develop techniques for analyzing illicit foreign explosives and to design more sensitive instrumentation for detecting contraband entering the United States. It recommended additional efforts to beef up the ability of law enforcement agencies to intercept suspicious communications and to enter buildings surreptitiously for counterterrorism investigations, all in accordance with legally proscribed procedures.

During the interim since that report was released, many fire departments and other emergency responders around the country have been vigorously preparing for a chemical or biological attack. As indicated below, federal agencies are attempting to anticipate and prevent such events that were not even seriously considered before the sarin assault in the Tokyo subway.

To further discourage hostile actions, the United States puts the squeeze to tender pressure points of countries that: 1) insist on sponsoring or supporting terrorism; 2) seek weapons of mass destruction; or 3) fail to take adequate steps to reduce narcotrafficking. Economic sanctions, political punishment, and even broader international blacklisting are discriminately applied to produce enough pain to get some attention.

Economic sanctions, sometimes called the liberal's alternative to war, have been described as coercive economic measures to force a change in policies, or at least to demonstrate displeasure. They fall into several categories. They may include a partial or total trade embargo, freezing of assets in the United States, and denials of access to U.S. markets. They may involve restrictions on aircraft or ship traffic headed for the United States, bans on private U.S. investments in the target country, and the termination of foreign assistance.[19]

Punitive actions can be triggered by U.S. laws that are aimed at specific countries. For example, the *Iraq Sanctions Act* prohibits the export of many goods to Iraq and Iran, and the *Iran-Libya Sanctions Act* prohibits trade by U.S. firms with foreign companies that invest more than $40 million annually in Iran's or Libya's petroleum development.[20] Actions can be pegged both to presidential orders or decisions and to actions taken by individual agencies, such as the Department of Treasury or the Department of Defense, as long as they have the White House's blessing.[21]

Working with allies, the United States can seek UN-imposed economic sanctions or, as we have seen, can mobilize support for broad-based restrictions on exports of military and dual-use items. Threatened restrictions on multilateral foreign assistance and bilateral aid also have their uses in the multipart strategy to isolate rogue states from the advanced technologies and fat financial resources needed to bankroll their bad deeds. Of course, in some cases our allies have their own special relationships with long-time trading partners. When such differing political agendas clash with hard-line policies advocated by the United States, we must unilaterally follow our own instincts for self-preservation.

Unfortunately, sanctions and other pressure tactics are often used by Congress as political footballs without adequate forethought of the consequences. Unintended hardships on democratic forces or helpless populations within the target country can easily result, as heartless rulers pass on economic penalties to those least able to pay the price. Nevertheless, sanctions remain a popular way for American politicians to demonstrate their outrage against the acts of governments that threaten international security.

As an extreme case, the Iraqi government claims that economic sanctions have resulted in the deaths of more than a half million children since 1991. Iraqi children are indeed currently suffering; but the Iraqi government cannot simply pass the blame to other countries. While the sanctions impose restrictions on oil sales, they have failed to curtail the continued high lifestyle of the ruling elite and the continued preparations for military action. Not that sanctions on Iraq should be lifted, but the Iraq case points out the dilemmas in trying to force change in ways that end up punishing people who cannot be held responsible for the intolerable situation.[22]

In terms of political punishment, the United States exercises a number of penalties to censure countries that risk international security; i.e., expulsion of irresponsible states from international organizations, severing of diplomatic relations, and expulsion of foreign students from the United States. Again, such actions are most effective when taken in lock step with other like-minded governments.

The United States has identified seven governments as being "terrorist" and special targets for concerted international action to minimize their threat to the world community, namely, Iran, Iraq, North Korea, Libya, Sudan, Syria, and Cuba. These countries are under continuing scrutiny at many forums where political and economic relations are discussed. Afghanistan would also be on the list if the United States had formal relations with the government of that country, but we cannot formally blackball a government that we do not recognize, according to diplomatic logic.[23]

The legislation calling for the Clinton administration to formally designate terrorist organizations, with attendant restrictions on their operations, required the Department of Justice to compile sufficient evidence to withstand court challenges.[24] With this caveat in mind, the attorney general and the secretary of treasury signed off on the thirty terrorist designates. Secretary Albright emphasized that other groups may be designated at any time. Having observed the secretary's modus operandi on a number of occasions, I think this can probably be construed as both a threat and a promise.

Turning to drug trafficking, we have already seen how the U.S. government's "certification" works as a mechanism for determining whether countries such as Mexico and Colombia are making adequate progress toward the reduction of narcotrafficking. "Decertification" triggers several automatic economic penalties, primarily related to denial of foreign assistance and of sugar exports, and raises the possibility of adjustments of tariffs and of limits on air and maritime traffic.

Overall, Congress has stocked its policy toolkit with an extensive array of devices with which to press foreign governments toward more responsible control over violence aimed at the United States and elsewhere. The willingness of the U.S. government to exert such power depends on many political and economic considerations. Perhaps the greatest weakness in the U.S. approach to blacklisting specific countries has been the heavy emphasis on extracting immediate political concessions while pushing aside the more tedious task of changing the environments where terrorism develops, a topic I will expand on later.

The delicate balance between hard-hitting condemnation and low-key diplomacy as techniques to change the behavior of an adversary was reflected in the hesitancy of the U.S. government to release a *White Paper* in 1997 deploring policies of Iran. While the analysis reportedly drew heavily on previous statements by Department of State officials at congressional hearings, the Secretary of State judged the document to be unnecessarily inflammatory and therefore did not approve its release. Presumably, the department had not given up on the possibility that quiet negotiations, rather than public bashing, could sometimes be more effective in encouraging Iran to modify policies that threaten American interests.[25] Within a few months, perhaps by sheer coincidence, the president of Iran was appealing over CNN for a more constructive relationship with the United States, initially emphasizing exchanges among private groups—wrestlers, scholars, artists, and the like—as a way of easing the tension.

The U.S. scorecard for punitive measures is not without its negative column, particularly as it has an impact on American business. It has been estimated that unilateral export sanctions in effect in 1997, for counterterrorism and also for other reasons, cost U.S. firms $15–20 billion, with over two hundred thousand American jobs sacrificed on the altar of economic restrictions. In addition, foreign companies steer clear of U.S. firms when placing orders for components for equipment headed to countries placed off limits by the U.S. government. U.S. sanctions intended to prevent Western firms from doing business with rogue states also may damage longer-term trading relations with close allies.[26]

One example of the consequences of sanctions imposed without multilateral support is the U.S. effort to reform Burma. Burma is ruled by a repressive dictatorship and fingered as the major worldwide source of heroin. While U.S. sanctions seem to have little effect on the Burmese government, they prevented the American company Unocal from participating in the development of offshore gas deposits. They forced Texaco to sell its interest in another gas field to a British company. The Burma experience also discourages these companies from looking for other opportunities in Asia where sanctions might later be imposed, thereby giving an edge to French and British competitors. Meanwhile, the state of Massachusetts has joined the struggle to pressure Burma to change its ways by banning purchases by state agencies from companies with investments in Burma. Even Harvard University reportedly turned a softdrink supplier away from its campus for selling cola in Burma. Business interests are of course worried that in such cases the only effect of sanctions—whether undertaken in retaliation for drug trafficking, support of terrorism, religious

persecution, unfair labor standards, or environmental transgres-
sions—will be to reduce the presence of U.S. companies in global
markets.[27] Such situations remain a tough call, forcing governmental referees to question whether American commercial interests should supersede a solid stance against cruel dictatorships
well fed by the drug trade. Yet the costs must be measured against
national security concerns of the highest order.

Limiting trade and transfers of technology and encouraging allies to follow suit, freezing financial assets, and adopting other pressure tactics to isolate countries can be very important. Unless used
with great skill, they can cause major diplomatic problems and economic dislocations with little positive benefit. Congressman Lee
Hamilton, noting that the economic costs of unilateral sanctions are
high and that they also are usually ineffective, has urged that before
the president imposes a sanction, the government should conduct a
cost and benefit assessment of the proposed sanction.[28]

We turn again to the troublesome country of Iran. Its renegade regime has been described on Capitol Hill as a militant nation intent on using terror both at home and abroad to achieve its
aims. Uninterested in international norms, Iran seems scornful of
public opinion and committed to developing weapons of mass destruction. According to the State Department, Iran uses terror
through Hamas and other organizations to disrupt the Middle
East peace process, promote subversion of neighboring governments, and incite political violence in places as far away as South
America and the Far East.[29]

U.S. policy has been directed to changing Iranian behavior
through economic and political pressure while stymieing development of Iranian weapons capabilities. This ensemble of preven-
tions and retaliations has delivered mixed results. A key objective
is to limit financial and technical resources flowing to Iran so as to
reduce the likelihood of its terrorist program flexing its muscles
beyond its borders. At the same time, U.S. diplomats stress that
there is no intention to isolate Iran permanently nor to overthrow
the Iranian regime. The goal is to encourage or force Iran to aban-
don those policies that have made it an international pariah.[30]

Frustrations in containing Iran peaked in 1997, when German
courts branded Iran's top leadership with direct responsibility for

the gangland-style slaying of four Kurdish dissidents living in Berlin. According to the Department of State, this brought to more than sixty the number of such Iranian-inspired assassinations overseas. Coupled with growing suspicions over Iranian complicity in the killing of sixty American airmen in the al-Khobar apartment complex in Saudi Arabia and the continuing Iranian offer of reward money for the elimination of author Salman Rushdie, the Berlin incident persuaded almost all American political leaders that a new round of harsh retribution was the only way to deal with Iran.[31]

For a number of years, Iran has been at the top of the list of designated terrorist states. The aforementioned *Iran–Libya Sanctions Act* had successfully reduced the financial resources available to the regime. However, at the end of 1997, the French company Total was ready to work with Iran in expanding production of its oil reserves, thus demonstrating Total's differing priorities and its willingness to give up any potential business it might have otherwise received from U.S. companies.

As to modern weaponry, Iran is believed to have a clandestine nuclear weapons development program. Its chemical warfare program produced 1,000 tons of chemical agents in 1996. Iran is capable of producing many different kinds of biological agents, according to publicly available documents. It has imported Scud missiles, probably from China, as well as components and technology for producing its own longer-range missiles. And Iran is buying conventional arms, with Russian imports reaching $1 billion in 1997.[32]

With such an incriminating curriculum vitae, Iran has been the focus of a continuing full-court press imposed by Washington for several years. Both bilaterally and multilaterally, American diplomats try to persuade supplier governments, and particularly European governments, to adopt effective measures to ensure that neither they nor their companies assist Iran's programs, directly or indirectly. Unfortunately, since France and Russia, in particular, do not share our high level of concern over Iranian behavior, a consistent international stance has not been possible.

In the nuclear arena, U.S. officials claim credit for slowing Iran's march toward the nuclear bomb, by convincing a number of countries to avoid contacts with Iran in this area and by pressuring

Russia to forego proposed sales of a gas centrifuge plant for enriching uranium and a research reactor that would produce plutonium to Teheran. Also, through the International Atomic Energy Agency located in Vienna, international inspections help deter nuclear weapons programs while improving the odds of detecting clandestine research and experimentation.[33]

Turning to other types of weapons, the U.S. government believes that it has succeeded in largely choking off European chemical and equipment companies as a source of supply. On two occasions the United States applied sanctions against European suppliers of equipment. Not to be deterred, however, Iran quickly turned to China as a new vendor. Now the United States has China in its sights in an effort to derail Iran's race toward chemical weapons.

As for missiles, again China and Russia are primary targets of U.S. urgings of constraint.[34] In trying to block access to conventional weapons, thirty-two of the leading industrial countries have joined the United States in agreement to deny transfer of armaments to Iran. Unfortunately, Russia is not one of them, having found a new market in which to generate desperately needed cash. The military equipment ties between the two countries seem certain to expand, despite Western protests.

As already noted, in early 1998, the president of Iran proposed private exchange visits as a step toward normalization of United States–Iran relations. He vigorously denied Iranian ambitions to secure weapons of mass destruction in support of a terrorist agenda. However, given the overwhelming evidence of Iran's complicity in so many terrorist activities, coupled with the bitterness of many Americans over the Iranian seizure of American hostages for 444 days, the U.S. government is naturally cautious in crafting a response.[35]

Critical alliances in a combined nonproliferation, counterterrorism approach to Iran depend on cooperation in the first instance of European governments, including France and Germany, and to the extent feasible, the cooperation of the Russian and Chinese governments. However, until there is broader consensus as to the seriousness of the threat of weapons proliferation to Iran and to Iran's intentions with regard to global terrorism, strong concerted action remains elusive.[36]

The FBI, which historically has confined its efforts largely to combatting subversion *within* the United States, is now expanding its role to include deterrence of foreign-inspired assaults on American institutions. Indeed, concerns abound in the Department of State that the FBI is taking over too much of the diplomatic turf in executing its charge to investigate attacks on American citizens and properties wherever they take place.[37]

At the same time, tracking plots of extremist groups operating in the United States is a critical part of any effective strategy to counter international terrorism. In this hunt, the FBI clearly is the lead dog. In some cases, our home grown terrorists may be keyed on the same institutions as international terrorists, albeit for different reasons. Their targets are frequently government agencies that they perceive to be pursuing controversial policies, although bombings of abortion clinics have been among the most common acts of violence in recent times. However, possible future connections of American dissidents with Russian mafia groups or Italian organized crime syndicates with bases in the United States cannot be ruled out.[38]

As noted, the Department of Justice and the FBI are well aware of the evolution of domestic terrorism. (See Table 27.) With expanded capabilities of 2,600 agents now devoting their full-time energies to combating terrorism—primarily in the United States— the FBI certainly should stand a good chance of uncovering suspicious plots within our country.[39]

The FBI emphasizes both prevention of and response to criminal acts, whatever the reason. Agents play down their role in analyzing motives and attitudes. Yet the FBI has established two new centers, a Counterterrorism Center and a Computer Investigations and Infrastructure Threat Assessment Center, in an apparent attempt to understand motives and attitudes as well as the incidents themselves. The FBI has also enhanced its laboratory facilities with a hazardous materials response section. Deployed to the Atlanta Olympics Games, this group will undoubtedly maintain a heavy travel schedule as it crosses the country to large-scale events, ceremonies, and competitions where impending nuclear, chemical, or biological incidents lurk in the background.

In reminding us of the overlaps between terrorism and other types of criminal behavior, the FBI director points out the example

TABLE 27.   Domestic Terrorism Since the Oklahoma City Bombing

- *Midwest, 1994–95*: Members of the white supremacist Aryan Republican Army go on a seven-state crime spree, leaving behind pipe bombs as they rob twenty-two banks from Nebraska to Ohio.
- *Vernon, Oklahoma, November 1995*: A self-proclaimed prophet and leader of an Oklahoma militia is arrested while preparing a bombing spree against civil-rights offices, abortion clinics, welfare offices, and gay bars.
- *Spokane, Washington, April–July 1996*: Citing biblical law, three self-described "Phineas Priests" commit bank robberies and bomb offices of the daily *Spokesman-Review*, Planned Parenthood, and a local bank.
- *Atlanta, Georgia, July 1996–February 1997*: Pipe bombs explode at Centennial Olympic Park, an abortion clinic, and a gay bar, killing one and injuring more than 100 people. The so-called Army of God takes credit for the clinic and bar bombings.
- *Phoenix, Arizona, July 1996*: Federal agents arrest twelve members of the Viper Militia and seize over three hundred pounds of ammonium nitrate—a key ingredient of the Oklahoma City bomb—plus seventy automatic rifles, thousands of bullets, and two hundred blasting caps.
- *Clarksburg, West Virginia, October 1996*: Federal agents arrest Mountaineer Militia members for possession of explosives and for allegedly plotting to blow up the FBI's fingerprint facility, where two thousand people work. Authorities seize TNT, grenades, and C-4 plastic explosives.
- *Kalamazoo, Michigan, March 1997*: Federal agents arrest a local militia activist for allegedly giving eleven pipe bombs to a government informant and plotting to bomb government offices, armories, and a TV station.
- *Yuba City, California, April 1997*: A blast that shatters area windows leads police to 550 pounds of petrogel, a gelatin dynamite, allegedly stored by local militia activists. The explosives are enough to level three city blocks.
- *Wise County, Texas, April 1997*: Federal officials arrest four Ku Klux Klan members who had planned to blow up a natural-gas refinery and use the disaster as cover for an armored-car robbery.
- *Fort Hood, Texas, July 1997*: Convinced that army bases are training United Nations troops to stage a coup, an antigovernment group plans to attack Fort Hood on July 4. Authorities arrest seven people and seize machine guns and pipe bombs.

Source: Bureau of Alcohol, Tobacco, and Firearms, Federal Bureau of Investigation, and Southern Poverty Law Center; as reprinted by *U.S. News and World Report*, December 29, 1997/January 5, 1998, p. 24.

of cybercriminals who intrude into our electronic circuits; they include not only terrorists, but also white-collar criminals, economic espionage agents, organized crime groups, and foreign intelligence agents.[40] Meanwhile, the FBI is shoring up its defenses around its own facilities with new screening devices and other protective measures.[41]

Then, in stressing the importance of prosecution as a principal means of deterring violent political radicals from the United States or abroad, the attorney general has stated:

> Although efforts to locate international terrorists and obtain their return to the United States are often very protracted, the passage of time does not diminish the government's ardor for pursuing these international criminals. In one case, for example, custody of a defendant was obtained and his conviction achieved nineteen years after his terrorist acts. In another case, the perpetrator of a deadly 1985 air piracy in the Middle East was tried and convicted in the United States in 1996. During these past four years, the relentless efforts to apprehend such fugitives have resulted in the return to the United States of seven individuals on charges relating to highly deadly terrorist plots.[42]

The bringing to justice of seven criminals in four years is a start, but the notches etched into the Justice Department's belt will have to be more numerous in the years ahead if America is to be safe from international terrorism.

Against a background of skepticism voiced by U.S. business interests and computer techies, FBI Director Louis Freeh in September 1997 presented his case to the U.S. Senate. He asked for new legislation arming the FBI and state law enforcement agencies with software keys to decode encrypted messages sent over the Internet. He argued vigorously that while the Fourth Amendment protects privacy, it also provides for intrusions into privacy, in accordance with appropriate judicial procedures, when criminal activities are suspected. With terrorists and other criminals using the Internet with increasing regularity, the director was reacting to recent technological advances underlying "strong crypto"—encryption of electronic messages in ways that ensure scrambled messages will not be deciphered for hundreds of years. He presented letters from every significant association of American law enforcement officials in support of his case.[43]

Freeh's particular plea was for a new law requiring an "unlocking window" be built into the programs of encryption software produced, distributed, or used in the United States. Such a back door would provide an access port for law enforcement agencies in possession of the unlocking software to monitor transmissions as they are being made. The FBI Director appealed:

> What good is it to intercept the messages of terrorists pursuant to a court order if we cannot understand what we have intercepted?[44]

According to Freeh, wiretapping and electronic eavesdropping are the most valuable tools available to the FBI in its efforts to combat terrorism—to obtain foreign and domestic intelligence, to obtain evidence, and to avert harmful attacks. Even though terrorists may be aware that interception methods are in progress, they nevertheless cannot resist the temptation to use the convenience of the telephone and the Internet in plotting their acts and in reporting their successes. Neither Freeh nor the Congress was interested in changing the ground rules as to requirements for court orders; Freeh simply wanted the technical capability to read the messages once he had steam-opened the digitized envelopes pursuant to court orders.[45]

Other experts pointed out to Congress that encryption is increasingly used by drug cartels and terrorist groups in their electronic communications, and cracking the codes has been very important in investigating a number of cases—from the Aum Shinrikyo to the New York subway bomber. (See Table 28.) Such cases were broken through a variety of means including the guessing of passwords, discovering helpful software, and otherwise lucking out. Such good fortune is quickly evaporating as criminal organizations increase their sophistication, shore up their defenses, and strengthen their encryption programs.[46]

Complicating the debate over providing law enforcement agencies with the keys to deciphering are the interests of American companies, who have operations spanning many countries. While acknowledging the importance of court-authorized access to encrypted messages by U.S. agencies, multinational companies are concerned that other governments would misuse such authorities. According to American executives, some foreign governments routinely monitor operations of American-based firms in order to provide foreign companies with proprietary information. The American firms are hesitant to report such incidents to the U.S. government lest their foreign hosts take retaliatory steps. The revelations of such vulnerability of electronic systems can erode

TABLE 28. Examples of Terrorism Cases Involving Encryption

- Aum Shinrikyo released sarin in the Tokyo subway: Japanese authorities decoded their files after finding the key on a floppy disc. The files contained evidence that was crucial in uncovering plans to deploy weapons of mass destruction in Japan and the United States.
- Ramzi Yousef planned the World Trade Center and Manila Air Bombings: Encrypted files containing plans to blow up eleven U.S. planes were decoded and found useful in the investigation.
- Terrorist Attacks on Businesses: An unidentified terrorist group attacking businesses used encryption to conceal their messages. American authorities found the key on a hard drive, and the decoded messages were a great help in the investigations.
- New York Subway Bomber: Encrypted files were eventually broken and while not particularly useful in the trial, they led to other computer-based evidence.
- Cryptoviral Extortion: Encoded viruses were introduced into at least nine business systems in London. The viruses encrypted bank records and files, criminals then demanded $100 thousand for the key.
- Spanish Basque Encryption: The French police have been unable to decode the hard disk belonging to a member of the organization.
- Cali Cartel: Encryption is being used to conceal identities of personnel involved in their networks.

Source: Dorothy E. Denning and William E. Baugh, Jr., Encryption and Evolving Technologies: Tools of Organized Crime and Terrorism, National Strategy Information Center, Washington, D.C., July 1997.

the confidence of stockholders in the integrity of company operations. Therefore, American business is eager to use strong crypto to protect internal operations without taking a chance that a key in the wrong hands would compromise their industrial secrets. In short, *the need to balance security concerns with business interests and civil liberties within the context of technological realities is the problem.*[47]

Another complication arises from the variety of applications of strong crypto currently available. For example, Visa Corporation relies on secure communications in dealing with 500 million card holders. At times, they put through tens of thousand transactions *per minute*. Naturally they are not interested in further complicating an already complex encoding system. For every transaction, they obtain an electronic record of the parties involved and the amount of their transaction. They argue that since this information can be made available to law enforcement agencies if authorized by court orders, it is unnecessary to break into financial transactions as they take place.[48]

Meanwhile, software companies object to Freeh's "costly and unworkable" proposal. They argue that savvy computer users could deactivate the unlocking technology thereby allowing the transmission of uncrackable messages.[49] In addition, the software manufacturers repeatedly point out that if they are limited in their products, particularly their exports, competitors from abroad will eagerly give the international market what it wants.

Privacy advocates contend that there should be no rules restricting encryption software, arguing it is critical to protecting privacy in the information age. They conveniently overlook the FBI's promise to act only on the basis of court orders. Or perhaps they believe that poking around databases is a different level of intrusion as opposed to simply wiretapping conversations.[50]

Encryption policy must satisfy a range of interests: public safety, law and order, national security, economic competitiveness, and civil liberties. The FBI should be commended. It is on the right track in seeking to cope with a new type of technological challenge by applying well-accepted principles of access to suspected criminal activities. Let us hope Congress will continue to recognize the importance of this issue and will respond boldly and responsibly, allowing the U.S. government to begin persuading other countries to follow suit.

However, the wired constituency of liberal law professors, high-tech organizations, and civil liberty advocates will not give up easily in their battle for Internet security, preservation of constitutional rights, and protection of industrial secrets. To view the extent of their *Lobby for Legislation* strategy, simply go to their web site at (www.crypto.org/join.html).[51]

Moving from steps to prevent incidents to measures to limit the damage, I have repeatedly underscored the importance of local responders as linchpins in successful efforts to minimize the consequences of a real-life terrorist-inspired disaster. A few additional comments help complete the picture of the current status of our emergency response capability. It is the local firefighter, not the federal bureaucrat, who is trying on the HazMat suit, adjusting the gas mask, and testing the bomb detection device. And it is the ambulance drivers and hospital nurses who will be unexpectedly thrown into unfamiliar territory in the event of a terrorist attack.

12. An inter-agency training exercise in New York City tested the mettle of emergency workers outfitted in HazMat suits, November 9, 1997.

Fire departments are inevitably on the front lines. The U.S.
Fire Administration in Washington repeatedly warns:

> The risks faced in today's world go far beyond the usual ones
> associated with residential fires, vehicular accidents, or even
> hazardous materials incidents.[52]

After pointing out that such threats may be categorized as Biological, Nuclear, Incendiary, or Chemical Explosive (B-NICE), the
Fire Administration adds:

> Be aware of the possible presence of a secondary device intended
> to injure or kill you and other first responders. Often, these sec-
> ondary devices are referred to as sucker punch devices.[53]

Instruction manuals prepared for first responders handling
terrorist incidents emphasize the responsibility in dealing not
only with a deadly incident but also with a crime scene where
high priority is given to preservation of evidence. Is it realistic for
a responder in a HazMat suit to take notes on any holes, breaks, or
scratches that he causes and photograph every object before he
moves it?[54] Fortunately, our fire departments have always put
public safety first; it is appropriate they be aware of the importance of solving the crime, but I am sure this responsibility will
not slow down their primary mission of limiting the damage,
while watching for booby traps. Most fire departments clamor for
increased training, expanded budgets, more personnel, and better
equipment. The federal government has taken important financial
steps to help local responders mobilize additional shock troops.

Before 1999, Medical Strike Teams should be in place in
twenty-seven metropolitan areas throughout the country. By
2002, 100 cities will have such teams in reserve. The early experience of the Washington Metropolitan Medical Strike Team gives
some insight into how they will function and the problems they
will encounter.

Under the leadership of the fire chief of Arlington, Virginia,
the 129-member team includes certified paramedics and HazMat
personnel, certified police officers, communication technicians,
three physicians, a toxicologist, a public health specialist and a
registered nurse. These personnel, on standby status, have a targeted response time of 60–90 minutes in the metropolitan area;

many teams members are from local HazMat units. The team is in effect on-site when the first HazMat responders arrive.[55]

The team should be equipped to detect many types of nerve, choking, blister, and blood agents and radiological materials. However, current detection capabilities need to be upgraded, since the detectors are susceptible to false positives, reacting not only to chemical agents but also to antifreeze, perfumes, and body lotions. As of 1997, the team had no capability to detect biological agents; moreover, military facilities in the area said they would provide only very limited support in analyzing samples. Even imperfect detection devices are very expensive and beyond the team's budget. Worse, the chief considers bioagents to be the least expensive and most attractive to terrorists.[56]

Perhaps the most serious problem encountered in Washington has been the lack of interest of hospitals, in supporting the team. After contacting thirty hospitals, the chief concluded that they could handle fifty—yes, only fifty—patients. Their reasons were multifold: other pressing priorities; costs of setting up decontamination areas; costs of training staff; lack of appropriate pharmaceuticals; and lack of communication links between hospitals and emergency response units, for example. In short, the hospitals may well lock their doors. Even ambulance drivers may be more worried about contaminating their vehicles than transporting patients. Such apathy led the chief to suggest that the licensing of hospitals and ambulances be tied to their willingness and capabilities to respond to chemical, biological, or nuclear emergencies.[57]

Turning to on-scene treatment of victims, the first concern is that perhaps 10 percent of responders could become victims of gases emanating from the scene of an incident. There currently is no capability for decontaminating the scene, which could inhibit entry of additional responders who may be less well protected. Finally, widely heralded plans to rely on the military experience developed for protecting young, healthy soldiers may not be very helpful in treating eighty-year-old victims.[58]

Finally, two other warning signs may dampen the chief's enthusiasm in moving the program forward. Since the U.S. Environmental Protection Agency considers waste water from decontamination activities as a hazardous waste, special (and very expensive) procedures for catching runoff may be needed as well.

Also, the chief believes that no matter how much preparation he and his team commit to and whatever their level of accomplishment, he will win or lose on CNN, which will be the ultimate judge of the effectiveness of a response.[59]

The wide variety of actions by federal and local agencies in preparing for the worst must be loudly applauded. They have certainly made an impressive leap out of the starting gates, but the race has only begun. Terrorism is too cheap, too available, and too effective in causing psychological fear.

With each successful terrorist act, there will be accusations that the officials responsible for public safety were negligent in preventing deaths and destruction. In some cases, officials will be replaced. In most cases, new legislation will be enacted.

The greatest danger is that we resort to reacting to the event of the month, rather than adopting a long-term comprehensive and consistent counterterrorism policy that does not waver from its course. In time, such a policy could provide the framework both for reacting to ad hoc events and most important, for preventing terrorist acts from occurring in the first place.

Searching for Long-Term Solutions to Replace Quick Fixes

Look for every opportunity to hit at terrorists.
Don't warn, don't threaten, don't send any
signals—only strike.
Department of Defense counterterrorism
official, 1997

The current policy mindset which sees all terrorism as
politically motivated and violent behavior may limit our
ability to combat new and expanding forms of terrorism.
Congressional Research Service, 1997

U.S. intelligence is optimized for secrecy and does not
have adequate access to substantive and culturally
critical information available from open sources.
Former CIA clandestine case officer, 1997

*I*t is New Year's Eve. Unlike millions of others who have planned exotic trips and foreign vantage points from which to observe the arrival of another early benchmark in the new millennium, you are nestled in your recliner ringing in the year 2003 with Dick Clark. He is on your big screen television, extolling the virtues of the new Times Square.

Before the big ball drops, however, the late night news airs its annual retrospective of the highlights of the year past. The spliced film clips come together in their collage of bright colors; there's not much there that would qualify as nostalgic or uplifting. In fact, you cannot remember a period of time punctuated with so many terrible events.

- In early January, an epidemic of an unknown virus struck five thousand people in Miami, causing severe respiratory pain, occasional choking, and even vomiting. Three hundred eventually died. Fearful of infection of other patients and of contamination of receiving rooms from what turned out to be a powerful new flu virus, the Miami hospitals refused to accept victims; they were taken to a nearby Air Force base for emergency treatment. All had attended the Orange Bowl game eight days earlier. All had visited the newly air-conditioned refreshment pavilion.

- Several weeks later in the Hague, a bomb exploded on the roof of the Europol building. The bomb, coated with radioactive material, rendered the building uninhabitable and contaminated several nearby blocks. Authorities are convinced that it was a retaliatory measure by a conglomerate of Russian, Italian, and Colombian criminal organizations because of pressure on them by law enforcement agencies using intelligence analyses from the Hague.

- In May the press obtained secret intelligence reports indicating that Libya, with China's assistance, had acquired an arsenal of over two dozen missiles armed with chemical warheads with a capability to hit southern Europe. The French and Italian defense ministers were alarmed but unable to convince European counterparts of the need for immediate economic sanctions. Despite U.S. calls for a military show of force, NATO refused to become involved.

- During the summer, three large financial firms in Sao Paulo, Brazil, discovered that intruders had effectively destroyed their computer systems; they were forced to cease operations. Records and backup systems were also compromised. An investigation revealed that other companies had made regular payments for several years to Brazilian criminal organizations, which had combined extortion with money laundering operations in the Seychelles. The three Sao Paulo firms had refused to cooperate with the criminals.

• The Japanese press reported in September that one of the country's largest private security companies provided the hit men for a series of assassinations of prominent businessmen—in some cases the very people they had been hired to protect. Efforts to investigate the case were stymied by Japanese judicial procedures. Meanwhile, Japanese authorities had made arrangements to provide businessmen with off-duty police officers for a substantial fee.

• AmTrak suspended all services during the Thanksgiving holidays on the East Coast following the derailment of an express train in New Jersey traced to an improperly closed switch. Seventeen deaths and 250 injuries were reported. Computer tampering had rendered the entire control network inoperable, and it took three days for service to be resumed. Investigators suspect insider–outsider collusion, but no arrests have been made.

• A new synthetic drug combining the characteristics of cocaine and heroin, but more powerful and addictive than either, hit the streets of New York in December. There were widespread reports of fatal overdoses. Raids by the Drug Enforcement Administration of recently established chemical laboratories in Puerto Rico indicated that Mexican criminals had moved some of their synthetic drug activities eastward as the security along the U.S.–Mexican border tightened. These laboratories may be the source of the new drug.

• On Christmas Day, the German chancellor who had led the efforts to reestablish law and order and public security in Europe was assassinated by a car bomb with no trace of the perpetrator. It was a plastic bomb, apparently hidden in a briefcase. There were no signs of chemical markers that would reveal the origin of the plastic material.[1]

Sadly, this calendar of events is considered credible by experts on terrorism and organized crime.

As I have emphasized, the nature of the terrorist threat—and our perceptions of it—are undergoing great change. Although low-cost kidnappings and bombings have been the order of the day for decades, high-tech attacks on high numbers of people and on the nation's infrastructure are now on the screen. Primitive terrorism will persist, but it is being joined by the increasing likelihood of superterrorism which exploits advanced technologies in order to deliver massive damage.

For years, terrorism was perceived by most Americans as an issue deeply rooted in the Middle East with newer outcroppings in Ireland, Latin America, Algeria, and Southeast Asia—realms for analysis by foreign policy wonks and of only passing interest to the average citizen. Countless acts of state-sponsored terrorism and atrocities of independent groups on foreign soil, in contrast to the calm at home, led to a widespread belief that devastation would be confined to distant lands. Although statistics have consistently shown that American citizens and property have been favorite targets of terrorists, the incidents against Americans have almost always been abroad, with little attention paid to the possibility of hits in America, thus further fueling a sense of denial.[2] Now, American leaders recognize that while terrorism may still be concentrated overseas, the United States itself is slowly but steadily becoming a favorite stage for enacting violent political dramas. As an example of what may be ahead, Senator Bill Frith, a well-informed medical doctor from Tennessee, has predicted that within five years there will be a chemical or biological attack on Congress.[3]

As detailed earlier, many international treaties and U.S. laws to help contain terrorism are on the books; efforts by American law enforcement agencies to neutralize this pervasive venom are growing steadily. What then are the missing ingredients of a strong policy and of effective programs to deal with terrorism? Is it simply a matter of efficiently implementing the laws or are there major gaps in the legal framework? Are counterterrorism programs focused on the most serious problems or are they off target? In short, do we have our arms around the problem, or are we wrestling with merely one head of a terrorist hydra, as the others commit their mayhem well outside our grasp? Recognizing this relentless multiplier effect of the terrorist machine, I explore here some new devices with which to behead the hydra before it devours us.

To attack terrorism on its many battlefields, I'll first restate in general terms what I think is the nature and scope of the threat, today and in the future. As we have noted, the term "terrorism" is used so broadly that, at least in the eyes of the public, it covers un-

13. The mega-cities of the future, with teaming slums like this one in Lima, Peru, may provide fertile breeding grounds for terrorists.

acceptable acts ranging from military atrocities to brutal rapes and
murders by plundering renegades to heinous street crimes, regardless of perpetrators or motives. Meanwhile, legal experts
search for defining characteristics to zero in on terrorism and to
set it apart from other types of unacceptable behavior. They argue
that if the definition goes beyond the traditional concept of politically motivated violence, it becomes too broad and almost
meaningless; we and other countries will have many practical
difficulties in responding. After all, for years we have operated
within a well-ingrained set of parameters that dictate appropriate
organizational and policy responses. The lawyers add that other
laws and structures are also in place, domestically and internationally, to deal separately—although often imperfectly—with organized crime, drugs, money laundering, and proliferation of
weapons of mass destruction. As noted at the outset, the U.S. government has taken this legal advice and works from the following
definition:

> Terrorism is premeditated, politically motivated violence perpe-
> trated against noncombatant targets by subnational or clandes-
> tine agents, usually intended to influence an audience.[4]

In many cases, terrorists have indeed been subnational
groups, including groups sponsored by rogue states. If the states
themselves carry out acts of destruction abroad, then we are talk-
ing about war according to past practice. But the distinctions between acts of governments, acts of their contractors, and acts of
independent subnational groups are not so obvious, as we have
seen with regard to Iran. When there are acts of devastation of innocent populations within a country, how should they be classified? If, as alleged, the army is a party to the brutal slaughter of
villagers in Algeria, aren't government officials just as guilty of
terrorism (even though international lawyers may prefer the term
"oppression") as the dissidents of the Islamic Salvation Front who
wreak havoc among the population?[5] How should the actions of
the Serbian authorities in Kosova be classified? Who is sending a
political message to whom?

The definition of state terrorism has other twists and turns. If
a country provides a base of operations for terrorists, such as Sudan, or a haven for terrorist retirees, such as Costa Rica, are its

leaders culpable for aiding and abetting? If a government turns its back on the passage of dirty money and smuggles arms across its borders en route to terrorist groups, shouldn't the responsible government officials also be held accountable? Most experts believe that state-sponsored terrorism is defined by such acts. Unfortunately, the list of countries avoiding difficult antiterrorism steps also includes dozens of nations we call our friends.[6]

State Department spokesmen strongly advocate that the United Nations promptly seek broad international accord in branding acts of terrorism as criminal acts, with the clear understanding that the leaders of independent terrorist groups are international criminals. However, the U.S. government is not prepared to label as criminals those foreign government leaders known to promote similar acts of terrorism.[7]

Even if the U.S. government was inclined to do so, there is no protocol or system to force foreign officials to trial for criminal or terrorist acts. Pressure applied by the international community is our only tool, albeit often limited in impact, by which to retaliate against or to punish such behavior.

Policy adjustments are clearly needed. As a starter, any government leader who permits the use of chemical or biological material surely should be designated an international criminal. An appropriate international tribunal should be convened to try such cases. The sentence should then go far beyond international censure, representing a penalty commensurate with the criminal act itself.

Having said that, I recognize that the concept of creating such a "crossborders" legal entity represents no trivial undertaking. Technical problems pose many a roadblock. For instance: (a) Information on terrorism activity is often obtained by clandestine sources and/or intelligence agencies; any trial proceedings would necessitate the exposure of sources, revelations that intelligence agencies would certainly resist. (b) Questions of impartiality would undoubtedly come into play when it came to selecting judges who would hear any given case. Once their verdicts were pronounced, enforcement of the decisions might in reality be merely symbolic or advisory in nature, because they would depend on the international will to punish a rogue state or person. The Saddam Hussein case provides a vivid example of the problem. (c) If conviction is achieved, carrying out the punishment is another challenge. Some countries do not legally have capital punishment. Other punitive measures may vary because of rules of law and degrees of freedom and liberty. If extradition is involved, the problems escalate. Extradition is already a hot issue even when related to ordinary citizens. How much more difficult would it be to obtain jurisdiction over a head of state and how many nations would actually hand over an official or head of state to such an international tribunal?

In time, nevertheless, the definition of international crime and terrorism must be further extended to embrace and clarify the knotty issue of countries sponsoring acts short of war that take human lives. While the international community has steadfastly strengthened its resolve to protect human rights, it must now expand on those initiatives to protect the very right to human life itself from new and omnipresent dangers.

Sharp legal differences have been drawn to separate (1) terrorists as groups or individuals that seek *political change* and (2) organized crime organizations and drug cartels as enterprises driven by a thirst for *accumulation of wealth*. Conventional wisdom has dictated that terrorists use violence to achieve political goals, whereas criminal organizations employ violence more selectively and only when bribery or intimidation fail. Terrorists have been characterized as intent on overthrowing specific political systems, whereas members of organized crime have been considered as targeting only those institutions that stand in their greedy paths to financial payoffs.[8]

The current reality is that life is not so jigsaw-puzzle neat. Such distinctions that had existed are fading fast. A few terrorist and criminal organizations already rely on the same global infrastructures for their illegal ploys, take advantage of the same breakdowns in authority and enforcement in states under siege, and seek increasing shares of the fortunes generated from narcotrafficking and other crimes. Whether mercenaries are hired to do the bidding of drug lords or of terrorist kingpins, the hit teams share a single motive in employing violence—earning their financial keep.

We have seen in Colombia that hybrid organizations, with parents of both terrorist and organized crime pedigrees, are a stark reality and hold the government hostage while pursuing a combination of political and economic agendas. Colombia may be the extreme case, but not the only case. We should be especially concerned that new hybrids may attempt to impose their will on other nations, cities, or institutions by threatening to unleash weapons of mass destruction.[9]

In short, modern forces of violence are increasingly incestuous, whether driven by political or economic desires. Grabbing control of financial institutions can both bring home the cash and advance political ambitions. Many groups, of course, will retain narrow portfolios of objectives, targets, and methods; others are becoming conglomerates of causes.

Returning to the traditional definition that refers to noncombatant targets, must there be direct human targets for an attack to be classified as terrorism? Sabotage of electronic networks, power grids, and other elements of a nation's infrastructure may not be aimed at crippling specific individuals or groups. Still, they may prove to be alarmingly effective in the disintegration of institutions, thereby disrupting thousands, indeed millions, of lives. Also, we have seen that while many drug traffickers think they do not promote violence against individuals, their profits end up fueling human devastation on many fronts.

Finally, diplomats are constantly seized with determining whether a certain brand of terrorism is international or domestic, sometimes a tough call. Cells of Middle East terrorist groups have been uncovered in the United States.[10] American extremists have forged links with radical groups in Europe. Asian operatives have set up shop in California cities.[11] Meanwhile, the tools of the trade have gone international; communications, organized crime, deadly weapons, and American dollars, authentic and counterfeit, touch nearly every port of call around the world.[12] In short, it is often impossible to identify the country of origin of weaponry, knowledge, money, or summons to action that is employed by terrorists, thus rendering meaningless in many instances distinctions between international and domestic terrorism.

Partnerships among groups with different agendas and linkages across international borders provide a reason to continue worldwide massive efforts to create tight legal and organizational frameworks that can slowly strangle international terrorism, domestic terrorism, organized crime, money laundering, proliferation of the most dangerous weapons and weapons technologies, narcotrafficking, and governmental corruption. Despite the best intentions of federal agencies, however, many of our counterterrorism policies are embedded in the narrow niches of earlier times. They need both broadening and melding in order to meet this fortified enemy head on. Efforts to develop such countermeasures are still works in progress; individual federal agencies have many opportunities to join forces and forge public and private partnerships. Indeed, coordination and partnership are the keys. We don't need yet additional agencies to confuse the issues. Rather, as terrorism thrusts tentacles in many directions, we must in turn multiply opportunities for synergies in countering terrorism domestically and internationally.[13]

With superterrorism looming large on the horizon, a new definition of terrorism is in order. And how do I define superterrorism? *I consider superterrorism to be the committing of violent acts using advanced technological tools to cause massive damage to populations and/or to public and private support networks.* I include all forms of nuclear, chemical, and biological attacks except small chemical poisonings. I include the use of plastic explosives but not fertilizer bombs. And I include cybercrime designed to knock out security, economic, and emergency systems.

While traditional concepts of terrorism may have withstood the test of centuries, the new millennium breaks the mold and calls for a rapid retrofit. Policymakers and lawyers have their work cut out for them.

Policies to address terrorism sometimes collide with human rights and civil liberties at home and abroad. In considering the extent to which we are willing to adjust to intrusions on those liberties, it is useful to view terrorism as a zone that lies at some uncomfortable mid-range between peacetime living and wartime emergencies. In peacetime, we enjoy all the privileges of protection of citizens' rights resulting from decades of interpreting and amending the Constitution, whereas in wartime national security interests, at times, override such rights. In the ambiguous state between these two poles, both philosophical and practical questions arise.

In general, the more limitations on the freedom of action of the government, the more limited the protection from terrorism. The government certainly has a duty to take extraordinary steps to protect its citizens when terrorists are still in the plotting stages, before they strike. If law enforcement authorities become paralyzed by legal procedures and cannot respond to emergencies in the making, they will not be able to do their jobs.

Should terrorism erupt on a massive scale in this country, the public will rightfully demand aggressive police intervention. Might we not put aside "proper" procedures for apprehending suspects if a clock's ticking was leading to detonation of a hidden explosive? Might we not be more aggressive in arresting shady characters fleeing from a booby-trapped building to an airport or other escape route?

We have mentioned many threats to personal freedom that are now at our doorstep, and still other thorny questions raise the ire of civil liberty advocates. How can the U.S. government respond to Middle East threats without prejudicing the rights of all Muslims living in the United States? Can air travelers who submit personal information to the airlines in advance be placed on lists that exempt them from security screening procedures, or would this be a discriminatory practice that favors those who are on the lists?[14] Should routine surveillance of users of the World Wide Web be permitted as law enforcement agencies gather information to indict terrorists who are recruiting or plotting over the Internet?[15] How vigorous can political dissents and even threats of retribution be without crossing the line of illegal activity?

In some American cities, the movements of ordinary citizens are on screen more than twenty times per day as security surveillance cameras spread like kudzu both in public and private areas. Cameras are being joined by a dazzling array of other technologies to keep track of our whereabouts and even to conduct automated strip searches from a distance. Some welcome this concern for the public safety, while others bristle at the thought that Big Brother is invading their privacy.[16]

The reality of the terrorist threat will add new dimensions to discussions of the Bill of Rights in the years ahead. (See Table 29.) New incidents of violence will strengthen the case for more

TABLE 29.   Relevant Provisions of the Bill of Rights

Amendment I
Congress shall make no law respecting an establishment of religion, or prohibiting
the free exercise thereof; or abridging the freedom of speech, or of the press; or the
right of the people peaceably to assemble and to petition the government for a redress of grievances.

Amendment II
A well regulated militia, being necessary to the security of a free state, the right of
the people to keep and bear arms, shall not be infringed.

Amendment IV
The right of the people to be secure in their persons, houses, papers, and effects,
against unreasonable searches and seizures, shall not be violated, and no warrants
shall issue, but upon probable cause, supported by oath or affirmation, and particularly describing the place to be searched, and the persons or things to be seized.

Amendment V
No person shall be held to answer for a capital, or otherwise infamous crime, unless on a presentment or indictment of a grand jury, except in cases arising in the
land or naval forces, or in the militia, when in actual service in time of war or public danger.

stringent law enforcement. New laws will push the constitutional limits of protecting civil liberties. Even those who advocate a new type of War Powers Act to battle terrorism may well
pick up support. While most of us do not think we are engaged
in warfare against terrorism, those terrorists on the other side believe they are. Unfortunately, their ranks are swelling.

The freedoms of democracy, as we have known the concept,
may need modification to permit strong action for thwarting terrorists on our shores. In entering the public policy debates, Americans must balance idealism with reality and recognize that
terrorism is a national security issue. Our system of justice has been
sufficiently resilient to adjust to many other national security dangers of the past, and it may soon be put to the test again.

On the international front, debates over appropriate responses to terrorism begin with varying interpretations of the
Charter of the United Nations and of the UN Declaration on Human Rights and then weave their courses through dozens of international treaties and agreements of many types. How can the
international community take actions against suspected terrorists
while ensuring that they will be entitled to fair and just proce-

dures? When should a nation extradite a fugitive who claims political asylum?[17] What rights do nations have in pursuing fugitives in other countries?[18] These issues are debated in cities around the world every day.

Other countries will retain their own definitions of civil liberties. And to American survivors of terrorist attacks, the stringency of Israel's antiterrorism procedures and Malaysia's harsh drug laws, for example, may look quite attractive.[19]

Another point in the personal freedoms debate centers around the character and integrity of the government employees to whom we are willing to submit ourselves for scrutiny. Civil liberty advocates are rightly concerned over evidence that at times, even in the United States, government employees are corrupt; you simply can't trust them. With specific cases to support this position,[20] large segments of the American public see abuse of government power as the primary enemy.[21] Still, our law enforcement agencies are central to the struggle against terrorism in the United States and around the world. Thus, it is of the highest priority that the agencies at all levels put their houses in order.

From FBI agents to state troopers and from customs inspectors to municipal police, all public employees involved in law enforcement must be held to a higher standard of conduct than employees in any other endeavor. The army, navy, and air force have their own code of conduct, the *Uniform Code of Military Justice*, that supplements the legal framework by which the average citizen abides. Similarly, the law enforcement agencies should have a specific code for their employees which is even more stringent than either existing Civil Service or military regulations

Many manifestations of such a code, such as random drug testing, financial disclosure, and confidentiality guidelines, are already in effect in numerous agencies. However, these strictures should be expanded and given sharper teeth through a swifter and more severe system of penalties.

With a no-nonsense code in place that reduces the temptation for public servants entrusted with sensitive jobs to stray from moral and ethical behavior, law enforcement officials would be able to argue more effectively for broader powers to combat all types and degrees of criminal activity, including terrorism. And most important, doubting Americans eventually could be swayed toward trusting that such officials are above corruption and are working for the public good.

Of course there will always be those who distrust the government, particularly the federal government, because of an abhorrence of any such intrusion into their lives. We can only hope that the overwhelming majority of Americans will eventually be persuaded to consider government intervention pursuant to the prevention of terrorism on par with well-accepted security measures performed at airports. Such scrutiny and inquiry about belongings and activities have generally been conducted on a reasonable, unbiased, and nondiscriminatory basis. If all such interventions are done in this way—and perceived to be activities for the safety and protection of all—perhaps such expanded security measures can be put in place without upsetting many people.

We are slowly recognizing that the public and private sectors need stronger alliances on many fronts to effectively stave off terrorism, particularly superterrorism. While public institutions may be imperfect and private citizens are reluctant to become involved in such distasteful activities, there is no alternative. Private sector facilities are prime targets of attacks; owners and operators of the facilities know their vulnerabilities better than anyone. Private individuals are aware of their personal vulnerabilities. At the same time, they are the eyes and ears of the nation. The public agencies, entrusted to protect our nation's people and our assets, have a special responsibility to work constructively with the private sector in taking precautionary steps, while paying greater attention to the perspectives and advice of those outside government when developing counterterrorism policies and programs.

Some types of cooperation are already mandated by law. As previously noted, companies file reports with the Department of Commerce concerning exports of sensitive items. Firms report information to Washington and to state capitals on chemicals they produce or import. Financial institutions keep records of all transactions exceeding $10 thousand and, if one were to go by the book, individuals must inform the IRS every time they carry $10 thousand or more in cash out of the United States. Laws also require information of a more general nature. For example, financial institutions are required to report all "suspicious" transactions regardless of the amount. Chemical companies are to report activities that pose threats for health or safety.

Another type of cooperation is tied to rewards-for-information programs. Federal agencies offer sizeable rewards, some in the millions of dollars, for information concerning diversion of nuclear materials, plans to commit terrorist acts, and the whereabouts of fugitives. Sometimes the U.S. government hesitates to publicize payments of such awards in order to protect informants, but this technique of encouraging whistle blowing has been credited with a number of successes—for example, heading off catastrophes during the Persian Gulf War and tracking down terrorists who have fled abroad, such as the recently convicted assassin of two CIA employees in Langley, Virginia.[22]

However, the record in using rewards programs has been spotty. As early as 1955, Congress authorized payments of rewards for information concerning illicit trafficking in nuclear materials. Forty years later, the FBI had still not developed guidelines for implementation of this program. New legislation in 1994 gave additional authority to the Department of State to pay for information about illegal acquisition of both explosive devices and dangerous nuclear materials. In 1996, during debates on yet additional legislation concerning protection of nuclear material, Senator John Glenn expressed the hope that it would not take a catastrophic event to inspire a reexamination of how this potential tool against terrorism could be made real.[23]

Another attractive approach has not been fully explored, namely, appealing to the sense of responsibility of American organizations and individuals to provide information simply because it is the right thing to do. Just as banks should have a know-your-customer policy, companies and other organizations handling dangerous materials should know their customers. Most companies belong to professional or trade organizations; governmental education efforts to keep these organizations, and in turn their members, apprised of developments in the field of terrorism need greater emphasis. An example of responsiveness in this regard has been the efforts of the Fertilizer Institute to encourage its members to report suspicious purchases once it was approached by the U.S. government after the Oklahoma City fertilizer-bomb incident.[24] Other suggested approaches, including several previously mentioned, include the following:

- The retailers and distributors of chemicals that can be used in many types of explosive devices should be embraced as full partners by law enforcement agencies in keeping these chemicals out of dangerous hands.
- The owners and operators of power, water, transportation, and communication systems should be in continuous contact with federal and local authorities concerning the security of these vital systems.
- Importers of goods from well-known drug-producing or drug-transit countries should know the histories of their shipments, including the reliability of loaders and transporters along the entire shipment route, as the narcotraffickers attempt to hide drugs in shipments of textiles, furniture, toilets, and all manner of goods.
- Administrators of high-tech training programs in the United States, including science and engineering programs at American universities and corporate training programs.
- Multinational companies should be aware of terrorist movements in overseas areas where they sell their dual-use products and should show restraint in servicing customers of questionable integrity.

Turning to the neighborhoods of America, we should be alert to strange developments around us and be less hesitant to inform authorities of suspicious goings-on. Ten years ago, most of us would have objected to the notion that we should keep an eye on our neighbors. But now, when a knock on the door raises unprecedented apprehensions as to who is there and when many neighborhoods have been subjected to violent crimes of one sort or another, a nationwide neighborhood watch program is becoming an essential element of life in America. Local police departments should help us appreciate the types of crimes uncovered in similar neighborhoods, so that we can raise our antennas to such possibilities. Unexplained laboratory facilities and machine shops, possibilities. Unexplained laboratory facilities and machine shops, warehouses stocked with explosive materials, and libraries of do-it-yourself destruction manuals should raise red flags.

Whistle blowing to fend off terrorism is an honorable deed.

Every expert and every book on terrorism emphasizes the crucial role of intelligence in detecting terrorists before they strike and in helping to bring perpetrators to justice after their pernicious acts. The directors of the CIA and FBI repeatedly reassure Congress they are on top of the problem and that their expanded staffs are giving the topic the highest priority. They claim many successes in preventing horrendous incidents from occurring and in tracking down terrorists in the farthest corners of the earth.[25]

Although new on the job, in mid-1997 CIA director George Tenet announced to Congress his laudable goals of freezing terrorists early and preventing and excising their operations before danger turns to disaster. Then he presented a list of counterterrorism achievements of his agency in recent years. Even though he reached back to successes in the 1980s, his list was surprisingly short. This accounting included: (1) providing information that resulted in an unspecified number of denials of American visas; (2) supplying information that helped bring thirteen fugitives to justice; (3) disrupting the plot to blow up American airliners over the Pacific Ocean that we discussed previously; and (4) discovering the Iraqi preparations to assassinate former President George Bush when he visited Kuwait in 1993. Of course, these were important accomplishments, but as a ten-year scorecard, it is not impressive. We can only hope the CIA submitted a confidential report that is more current and more extensive.[26]

At the same time, it is unrealistic to expect that standard intelligence methods—eavesdropping, photographic satellites, reports from informants, observations by hardworking gumshoes, and analyses of media reports—are adequate to identify and defeat all foreign-based organizations eyeing America. The number and variety of known groups of concern, let alone the unknown groups, are simply too large. Missing just one could have enormous consequences. Thus, we must recognize that the emergence of superterrorism demands modifications both in mission and in scope or direction, the transition will not be easy. For the CIA, its historic modus operandi of secrecy and covert action makes the task even more difficult.

For example, Congress recently put the CIA to the test by requiring semiannual reports of the spread of dangerous technologies that feed rogue states. Unfortunately, the agency's initial response did not earn a passing grade. In a very short document, the agency reported a few cases of four countries (Russia, China, Germany, and North Korea) providing high-tech weapons and equipment to eight countries of concern (Egypt, India, Pakistan, Iran, Iraq, Libya, North Korea, Syria), pointing the finger primarily at China and Russia as the sources of the problem. Not that longer reports are necessarily better, but even an untrained observer could have prepared a more complete report using only press accounts.[27]

The CIA surely knows that transfers of "technology" involve much more than shipments of tangible equipment. Indeed, the most important aspect of technology is the *knowledge* to produce lethal products and weapons; the CIA should have identified several dozen countries of concern that are in the process of developing such capabilities, with supplier countries also numbering in the dozens. The biggest plus for the report was its public availability. It is likely that the CIA also has delivered to Congress its confidential assessments, but the penchant for secrecy does not help many outside the closed halls of government who have a stake in deterring terrorism. Six months later, the Secretary of Defense released his public report on the same subject, a report that was much more informative, proving not only that more information was available, but that the government itself could make it public.[28]

Putting the reports aside, the confidence exuded by the director of the CIA that his agency is on top of the problem may not be fully warranted. How did the New York subway bombers, who had lived on the fringes of terrorism in the Middle East, slip through the intelligence net in 1997, with only a last-minute tip saving a disaster in downtown Manhattan?[29] How confident can we be that more terrorists are not busy at work at this very moment in preparing to take out major facilities in the United States?

While the CIA reportedly devotes enormous resources to tracking down individual foreign-based criminals *after* the damage is done—which is extremely important from the point of view of deterrence—the more difficult task of anticipating who will strike next and where they will strike simply cannot receive too much attention. With increased terrorist activity around the world, this goal must be the order of the day. Albeit less dramatic than escorting fugitives back to courtrooms of Virginia from hide-aways in Pakistan, *prevention* of terrorism is the only cure.

Of course, the problem is incredibly complex. Oppressed people in many countries view the U.S. government as the incarnation of evil and greed. The Iranian government calls the United States "the great Satan," and Baghdad accuses Washington of plotting to starve the future generation of Iraqis. Such accusations are quickly adopted as rallying cries by terrorist organizations. They do not narrow their sights to military installations or even to heavily populated areas; they may be targeting any community, institution, or structure in the United States or any symbol of American interests abroad. They may be targeting American officials, business executives, or ordinary citizens, hence the sometimes random nature of their attacks. For the drug lords, of course, the bull's eye may be of a different sort: simply kids on the streets that are potential consumers with long-term brand loyalty.

New concepts of intelligence must go beyond traditional data collection and reporting procedures. The CIA should continue to respond to information-collection assignments prepared by government agencies that are "users" of intelligence and then prepare intelligence briefings for these agencies and indeed for the president himself. But their constituency has expanded. "Customers" for intelligence are no longer only the president, his advisers, and other federal government officials. A much broader range of local law enforcement officials, first responders, and individuals most likely to face the initial fallout of terrorist attacks have their own needs for accurate and up-to-date reports. Most of these people do not have security clearances for secret briefings. The challenge to the CIA is to serve the public, while at the same time taking care of the primary tasks of informing the president and other government officials and agencies. The following suggestions could help address new types of threats while bridging the current gap that separates intelligence collectors from the public—without de-grading intelligence support for the president.

1. Develop New Templates for Intelligence Analysis
of Easily Hidden Terrorist Plots

In evaluating whether weapons of mass destruction are on the loose, intelligence agencies have long relied on "indicators," on-the-ground developments that warn of rogue states acquiring the capability to use weapons of mass destruction for military purposes. If weapons are to be militarily effective, the states must operate on a rather large scale, particularly at the stage of testing the weapons. It is difficult to contain secrets known to too many people. The skilled humanpower and facilities required to assemble nuclear, chemical, or biological weapons have many telltale signs that can be detected by satellites, passersby, and professional informants. Nearby residents raise questions when there are new government structures, new employment opportunities, and new purchases arriving by land, air, or sea.

When it comes to terrorist groups that are interested not in arming themselves for military battle but only in disbursing dangerous materials of mass destruction without the need for high levels of efficiency, disguise is much easier. Careful review of the undetected approaches of the Aum Shinrikyo is an obvious starting point for identifying these more subtle clues of their intentions. Of course, this does not mean that the next high-tech terrorist group will opt for the same type of itinerary: traveling to Russia in search of weaponry, organizing a testing program in Australia, having their own small laboratories, and attracting tens of thousands of followers and assets exceeding $1 billion. Nor will the next group necessarily have publication services or international communications networks such as those the Aum Shinrikyo operated. But still there are basic requirements for skilled people, financial resources, and dangerous materials that would be necessary for any group to pose a significant threat.[30]

The Department of Defense reportedly uses a system of designated "threat levels" when tracking conventional or high-tech terrorists—levels determined by assessing the following criteria: Have the perpetrators succeeded in entering the United States? Does evidence point to their intention to carry out destructive acts? Do they demonstrate the necessary capability, and/or have a track record in carrying out terrorist activities? Such classifications must draw on indicators of intentions and capabilities.[31] However, as long as the intelligence agencies are isolated from the larger world of public information enjoyed by the private sector, by inclination or budgetary constraint, their ability to appreciate the importance of some subtle indicators and predict future events will be limited too.[32]

2. Make Stronger Commitments to Use All Potential Sources of Information

The intelligence agencies need to employ the latest information technologies to help separate reliable tips from misleading reports. They should rely more heavily than ever before on the expertise of others who have critical insights concerning terrorist activities. As long as the CIA, for example, clings to its belief that covert operations are the lifeblood of its operations, secrecy will continue to be both an asset and a liability. Some CIA analysts are reluctant to use authoritative documents available on the Internet because to them unclassified documents are not completely reliable.[33] In continuing to promote a culture of secrecy, the CIA director underscores: "At the end of the day, this is an espionage organization."[34]

A few intelligence officials are calling for a new emphasis on *open sources*. Open sources can include everything from newspapers, trade and industry journals, and ethnic and religious publications to speeches and papers by officials, scholars, clergy, and the like. Such sources can sometimes provide key information not to be ignored or disregarded. However, the intelligence community does not use all sources, although they have increased their reliance on academics, private consultants, and a corps of retired intelligence officers, all of whom have top secret clearance and are reimbursed for their contributions. While a step forward, such an approach continues the reliance on only "old-boy" networks, important but limited in scope.[35]

As already noted, the President, his advisers, and their staffs are obviously of supreme importance to the intelligence agencies. The Washington law enforcement agencies are key consumers. But Washington is a long way from Centennial Park in Atlanta, or the Murrah building in Oklahoma City, or the subway controllers in

New York. Time is of the essence in heading off terrorists and, failing that, in responding when they set off their devices of doom.

Thus, another way to expand information sources is for both the international and domestic intelligence agencies in Washington to form external advisory committees that operate on an unclassified basis, simulating the way most of us work. Such committees should include officials of local governments, local responders, industrial security specialists, financial specialists, academics with no paid ties to intelligence, politicians, and ordinary citizens. Their interactions with the agencies could clarify both how the private sector can help in providing reliable tips and how organizations outside Washington can more effectively benefit from the intelligence reports prepared by government.[36] The intelligence agencies have dedicated people; now let them have access to equally dedicated professionals beyond the Washington beltway, ready to help in roles other than as paid consultants.

3. Do Not Hesitate to Use Unconventional Methods when Warranted to Undermine Individual Terrorists; Thwart Planned Attacks; and Punish Irresponsible Governments

While assassinations have been ruled out by the U.S. government, bringing criminals to the United States for trial and sabotaging weapon caches of terrorist groups are not, if the countries where they hide do not object. As was demonstrated in the American bombing of selected targets in Libya in 1986 and in Iraq in 1993, the president can and should go beyond the norms of customary law when the national interest so dictates.[37]

Reinforcing the president's authority to get tough with terrorists is a recent statement reflecting a consensus within the U.S. Senate:

> If evidence suggests beyond a clear and reasonable doubt that an act of hostility against any U.S. citizen was a terrorist act sponsored, organized, condoned or directed by any nation, then a state of war should be considered to exist between the United States and that nation.[38]

Of course, extreme caution must be practiced to avoid pulling the United States into all-out warfare—unless the security stakes should warrant that level of response. Regardless, terrorists should be made aware of the united front among our government representatives.

With a war chest of $26 billion per year, the intelligence agencies have enormous staffs. They have the world's most powerful computers. They have reward money. In short, they have nearly carte blanche as to resources with which to interdict terrorists. By partnering with other Americans who have equally sensitive ears to the ground, the intelligence agencies can add substantial bang to the buck budgeted for keeping the terrorists at bay. Such a coalition is the kind of networking that will uncover terrorist lairs before they can mobilize their forces to wreak havoc in Tokyo, New York, and other cities of the world.

As we have repeatedly seen, ethnic clashes, religious-based terrorism, nuclear smuggling, drug trafficking, and money laundering are on the rise around the globe, particularly in Europe. Bombings in London, Paris, and Berlin have been rooted in domestic and international discontent of terrorist-oriented groups. Each seizure of nuclear material of any sort, no matter how harmless, raises the specter of inroads by Russian mafia groups. But drug trafficking and money laundering are too often passed off as business as usual on the continent. While the United States cannot be expected to check terrorism that erupts in Europe and may even find its way across the ocean, we can wield tremendous influence on holding the line through an organization which has spent the last seven years searching for new missions—NATO.

Since the splintering of the Soviet Union at the end of 1991, the NATO planners in Brussels and Washington have had great difficulty constructing realistic threat scenarios for NATO attention. U.S. officials acknowledge that there are no near-term "enemies," but argue that NATO will preserve stability in Europe. When reminded that for three years NATO was impotent with regard to Bosnia and that Italy, not NATO, responded to turmoil in Albania, they agree that NATO must do better to retain its credibility.[39]

If maintaining stability in Europe and on its periphery is the name of NATO's new game, then what better issue for NATO than the threat of international terrorism? In the near term, terrorists will not tumble governments, but they can cause considerable damage. As the Basques demonstrated in Spain, they surely will divert the attention of many countries away from more productive endeavors.

According to Article 5 of the North Atlantic Treaty of 1949, "The Parties agree that an armed attack against one or more of them in Europe or North America shall be considered an attack against them all." It will be difficult for NATO countries to interpret terrorist assaults within the scope of Article 5, or even an amended Article 5. In such a scenario, the governments of England and Ireland would be under great international pressure to ensure that the problems in Northern Ireland have been resolved, with NATO countries obliged to assist if requested. The French would have to deal more harshly with Algerian infiltrators, relying on the international community for both political and physical support if need be. And the most important development for Americans, an internationally based attack within the United States, would require a vigorous response by many European allies against terrorists seeking sanctuaries.

As a starting point, NATO could certainly craft a new article, placing counterterrorism on its security agenda. Closer integration of intelligence functions is a logical first step, followed by joint training and response exercises. In time, counterterrorism could become a major NATO undertaking as the European countries recognize that the NATO military infrastructure can provide important support for law enforcement agencies charged with stopping terrorists in their tracks.

If we begin to liken the battle against terrorism to war, NATO's commander for southern Europe seems sufficiently tuned in. He argues that the best prevention of the awful cost of war is a robust posture that is forward-engaged to confront a crisis in its earliest stage at the source.[40] If we can forward deploy troops against a yet-to-be defined foreign aggressor with designs on Europe, we can certainly exert enforcement muscle on the encampments of international terrorists in Europe.

NATO engagement in traditional counterterrorist activities may be a hard sell. Other international organizations, let alone national authorities, will object that NATO is moving onto their turf. NATO can, however, act quickly and decisively while other European-based organizations spend years debating definitions and international protocol. Despite overlaps and turf intrusions, the counterterrorism mission seems made to order for an action-oriented and well-funded organization of NATO's character. Although the most expansive role in counterterrorism should still be played by law enforcement agencies of various countries, NATO could certainly throw its weight behind these efforts.

Thwarting nuclear, chemical, and biological terrorism is a responsibility one would think military leaders of NATO might embrace. Yet when I suggested this approach to officials in Brussels in 1997, they stated that they had only reluctantly agreed to support a series of scientific workshops on bioterrorism, a topic well outside their priorities.[41]

In short, the U.S. government seems to be largely ignoring an important tool at its disposal to confront terrorists in Europe.

Finally, we cannot fail to address international terrorism born in the exploding populations of the developing countries. Terrorism's sprouts are manifold, entwined with almost every aspect of economic and social development. Simple approaches will not solve the problems; it will take a well thought-out, long-term strategy and an equally firm commitment by the United States and its allies to make a significant difference.

Historically, much of terrorism's fuel has been pumped from the wells of ethnic and religious conflict. This will be the most difficult facet to fight. However, modern mutations of terrorism are also inextricably linked to overpopulation, food shortages, and environmental devastation. Reducing these problems should be an early target in our sights. Ethnic groups living below the survival line will not remain passive as they watch neighbors enjoy a better life. Helping deprived peoples climb out of their holes of poverty will be an important agenda item, especially if it can relieve some of the tension inherent in the conflicts mentioned earlier. We must keep in mind also that whether these problems manifest themselves in terrorist acts or in criminal behavior, the two are inextricably entwined. Efforts to prevent one activity will also have a preventive effect on the other.

At the same time, worldwide appetites for advanced technologies must be carefully managed. Many sources can fuel terrorism with money, arms, technology, and equipment. And, consequently, containment of criminal behavior with its dangerous new partnerships with terrorists is an essential ingredient of successful counterterrorism efforts.

In preparing long-term programs for improving the lives of people in developing countries, we must customize them to be country-specific or region-specific. In Asia and Latin America, for example, the seeds of terrorism are increasingly incubated in the slums of megacities, in pockets of rural poverty, and in fields of poppies and coca leaves where farmers who would be destitute without this crop. In Africa, hordes of refugees have descended on regions of abject poverty, while mercenaries of South Africa stand ready to do the bidding of paying governments and arms and drugs flow through the ports of Nigeria. In the Middle East, weapons of mass destruction are at the top of every worry list, and the very existence of Israel continues to be a lightning rod for extremists in neighboring countries.

Much violence can be traced to the struggle between haves and have-nots with the boundaries between war and terrorism increasingly blurred. But of all the U.S. congressional priorities, foreign aid has been at the bottom of the list for many years. Why should the United States pay the bill for solving the problems of the poor countries when corruption, hostility, and rivalries characterize the actions of the leaders of these countries? Why should the United States invest in upgrading life in distant megacities when much of our own population does not have access to adequate education, consistent employment, or personal safety? This is a common refrain of congressional leaders. And they certainly have a point.

On the other hand, for years the Bush and Clinton administrations have argued that foreign aid investments are good for the United States. They promote the concepts of democracy and private investment essential for countries to become responsible members of the international community. Economic growth south of our border counters incentives for illegal immigration into the United States, reduces the likelihood that contagious infections and environmental calamities will reach the United States, and opens opportunities for American companies to invest abroad, so argue our presidents.

Our foreign aid programs have repeatedly been given credit for contributing to the emergence of democratic governments. Now with comparable vigor, our programs need to address job creation for the young would be entrepreneurs living in environments that currently offer few choices. Those young people with the guile to thrive in the barrios and ghettos are precisely the type of youngsters who, given the chance, could be successful in industry and commerce—rather than in the various criminal empires of the world.

We simply have not yet seriously factored into our foreign aid portfolio recognition of the necessity to dampen the spread of terrorism toward the United States. Failing that, we have omitted one of the most powerful arguments that could be mustered in raising the priority of foreign aid. But I am not proposing to right this problem, since with few exceptions foreign aid initiatives are immediately tabled once they reach Capitol Hill. I do propose a new approach, however.

During the past fifty years, the United States spent trillions of dollars to deter a nuclear attack by the Soviet Union while also holding the Soviet conventional arms powerhouse at bay. With the end of the Cold War, Senators Sam Nunn and Richard Lugar provided the leadership that enabled us to launch a program for preventing, albeit imperfectly, weapons of mass destruction technologies from flowing out of the former Soviet Union. The world is indeed a safer place as the result of their efforts, which led to the Cooperative Threat Reduction (CTR) program discussed earlier.[42] These two senators, together with Senator Pete Domenici, have also had the foresight to recently append to the CTR Program funds for supporting preparations to respond to terrorist attacks in the United States.

Now we should consider a CTR_2 program—a Cooperative Terrorism Reduction initiative. I underscore the word *cooperative*—not foreign aid but a program that benefits both sides and depends on contributions from both sides. This program, would operate in tandem with the many quick fixes we currently support to thwart terrorism in the very near term. The CTR_2 program would be directed to creating major changes in those countries of greatest concern—perhaps beginning on a pilot basis with countries where the task seems less difficult than others, such as the Philippines and Peru. In time, the program could target countries with more complex dynamics, such as Colombia, Algeria, and Pakistan. The funding levels must be substantial—billions of dollars annually to start—since the task is to promote legitimate economic development as a real alternative to illegal commerce in dangerous commodities.

Of course, it is critical that such a program be carefully coordinated within the antiterrorism strategies currently targeted on these same countries. With much more financial support at stake than today, the participating countries would have stronger incentives to commit to fundamental reforms—an all out attack on government corruption, systematic eradication of terrorist activities based within the country, prohibitions on imports of sophisticated weaponry, elimination of narcotrafficking, and stiff penalties on financial institutions that engage in money laundering.

To win the Cold War, we never shirked at the price tag of trillions of dollars. To win a war that is already heating up, we should be prepared to ante up some significant portion of that during the next decade, recognizing that in ten years we will have only fought the early battles in a century-long campaign.

There is some experience on which to build. For years, the United States has supported security assistance programs in many strategically located countries. The most successful example has been American support for Israel. Despite its seemingly impossible security problems, the country has been able to develop an impressive industrial and agricultural base while absorbing refugees from many countries, greatly increasing the population. Realizing that Israel is unique in many ways and perhaps may only serve as an appropriate model for states where a democratically elected representative government exists and functions, we should still be able to adopt or modify some examples of successful sustained development achieved in the face of constant security threats.

A CTR_2 program should operate with several cardinal principles. First, as previously mentioned, job creation would be the cornerstone of the program. Secondly, self-empowerment of local organizations committed to improving the standard of living would provide constructive outlets for both leadership skills and aggressiveness that might otherwise end up promoting criminal activity that fuels terrorism. Next, complete openness, often called "transparency," in all actions involving advanced technologies is essential. For example, development of medical, veterinary, and pesticides capabilities should be strongly supported in exchange for transparency of such activities that have implications for bioterrorism and chemoterrorism. Finally, imports of armaments— beyond the essential self-defense needs appropriate to any state including arms from the United States and its traditional allies as well as smuggled arms—should be minimized or banned as a precondition for participation in the program.

Of course, the devil is in the details. Difficulties in implementing effective programs have almost always been the Achilles heel of the Agency for International Development (AID). Indeed, AID should not be the implementer of the program; the Department of State is a far better locus for sustaining the cooperative character of the program. Also, a new breed of American program managers is essential. No longer should American experts be unilaterally designing the specifics of individual projects in other countries. Team players should work in partnership with recipient organizations to devise the blueprints for effective programs. No longer should U.S. contributions be measured in terms of numbers of advisers—advisers who are often unwanted, unneeded, and unheeded. No longer should the quest for quick fixes of long-term problems be the strategy for solving deep-rooted problems. In short, maximum responsibility and accountability should be shifted to recipient governments. At the first sign of corruption in the use of funds, the programs should cease without debate. Only then would the countries receive the message that when large sums of money are at stake, we are serious that rules will not be broken.

Substantial financial contributions to such a program would provide the U.S. government with enormous leverage in ensuring real commitments of governments to sustain programs that create jobs while taking steps to cleanse government departments of corruption. In the long run, the investment would seem small in comparison to the costs of defending American interests on an ad hoc basis by reacting to the constant eruptions of terrorists whose few alternatives inspire rage and retaliation.

As noted at the outset, recording the history of violence is to recount the entire history of civilization. At the dawn of the third millennium, however, our technological advances—wrought by the wrong hands—threaten to outdo anything on the terrorist timeline that has ever come before; superterrorism has the potential to eradicate civilization as we know it.

Several conclusions concerning the future responses of the U.S. government and the American people to the growth of terrorism seem clear.

- While the Cold War of nuclear intimidation lasted fifty years, we should prepare for a much longer battle to prevent high-tech superterrorism at all costs and to keep conventional terrorism at a tolerable level.
- It is a predictable impossibility to eliminate all the direct motives that drive individuals and groups to adopt terrorism as their modus operandi for addressing their grievances. But progress can be made by confronting the underlying motivations in a major way; only then will there be a real possibility to reduce the frequency and magnitude of future incidents.
- Our international program commitments must be measured not in months, not in years, but in decades. Discernible results will be slow to come. Unless a cooperative program is sustained, it probably should not be undertaken in the first place.
- No single government agency or program will in and of itself cause the transformation of problems faced at home or abroad, but aggregated programs can be very significant.
- Civil liberties are important for our democratic society; the time has arrived, however, to reconfigure some aspects of our concept of democracy given the violence that is on our doorstep and indeed has already hit us where we live. Minor finetuning now can reduce the need for major changes in lifestyle in the future.

Terrorism at some level will always be part of life on the planet. Major crimps in the supply lines that feed the terrorist frenzy can reduce the frequency and intensity of destructive acts. Economic and social alternatives may also minimize the motivation to continue such madness.

Epilogue

*Deal with terrorism where you can, but recognize that it
is part of the cost of doing business in free societies.*
Department of Defense Counterterrorism
Official, 1997

*If they don't watch it, the United States will be at the
top of the hit list of all Moslem states.*
Nizar Hamdoon, Iraqi Ambassador
to the United Nations, 1997

*Traditional notions of sovereignty, national security,
and warfare will be undermined by the year 2020 when
the whole world is wired and e-cash is the norm.*
Clinton Administration spokesperson, 1996

"*O*ur personal, community, and national security depend upon our policies at home and abroad. We cannot advance the common good at home without also advancing the common good around the world. We cannot reduce the threats to our people without reducing threats to the world beyond our borders.

"We prevented attacks on the United Nations and the Holland Tunnel in New York. We thwarted an attempt to bomb American passenger planes from the skies over the Pacific. We convicted those responsible for the World Trade Center bombing.

"While we can defeat terrorists, it will be a long time before we defeat terrorism. America will remain a target because we are uniquely present in the world, because we act to advance peace and democracy, because we have taken a tougher stand against terrorists, and because we are the most open society on Earth. To curtail the freedom that is our birth right would be to give terrorism a victory it must not and will not have.[1]"

With these confident words, delivered at George Washington University in August 1996, President Clinton set a tone for U.S. policy on terrorism that all Americans can support. But translating his oratorical punch into the kinds of powerful programs that will prevail over superterrorism is one of the greatest challenges of the next century. Rogue states may well become increasingly reckless with weapons of mass destruction. Terrorist groups may multiply in number and increase their firepower. Resumes of unaffiliated terrorists and hired hands grow more diverse—from computer programmers to rocket scientists. Global trends toward open borders, expanded commerce, and e-cash greatly complicate efforts to hold terrorism in check.[2]

Overall, the cauldrons of terrorism throughout the world are steaming with pent-up violence as they reflect a planet of limited resources tormented by a growing population. New alliances are being formed, bent on damaging and extorting governments and institutions that symbolize political and economic injustices. They feature well-organized criminal structures, skilled managers with access to deep financial pockets, and high-tech specialists intrigued by new opportunities to exhibit their skills.

Throughout this book, I have addressed many "credible" threats of terrorism. Are there even more extreme dangers ahead? Should we brace for a blast from the deliberate explosion of a tanker, loaded with 125,000 cubic meters of liquified natural gas, anchored in one of our harbors, an explosion that has been likened to the blast from a small nuclear bomb?[3] Should we heed warnings that genetic engineers may be able to create viruses that attack only certain ethnic groups—Caucasian, Black, Asian, or Jewish?[4] Is Secretary of Defense Cohen realistic in warning of possible triggering of earthquakes or volcanoes by terrorists with electromagnetic pulse machines?[5]

By sheer coincidence, the magazine *Wired* selected the year 2002 for its doomsday scenarios, the same time period I have chosen for my closing chapter. This respected journal suggested cybotage leading to phone lines going dead from San Diego to Portland; jets colliding after radar screens freeze; Alaskan oil pipelines springing leaks following implantation of a computer virus disabling the leak detection system; a microwave-activated bomb exploding in the Pentagon; and control malfunctions triggering a nuclear power plant accident, with eventual discovery of the perpetrators on the coast of the Black Sea.[6]

As we have seen, barriers of international and national laws have been erected and policies of deterrence and defense are in place to repel more conventional threats. In addition, nations must share information on renegade operations as never before and cooperate in running down terrorists. Our public and private sectors must work hand-in-glove to resist an increasingly aggressive adversary thrusting its tentacles into American society. Physical protection of facilities and people coupled with warning and response systems to deal with impending disaster are the final lines of defense. As each country and each community erects its shields according to its own specifications, terrorists will identify the most vulnerable barriers that protect pockets of power or wealth. Until the most vulnerable are no longer easily penetrable, terrorist predators will continue to feed on society.

Once a bomb has been detonated, a hostage seized, an electronic network disrupted, or a disease virus released, the response of the nation and the community dedicated to protecting its citizens should be immediate, with pursuit of the perpetrators relentless. When those responsible are captured, punishment must be stern and unyielding. Only when plotting terrorists know that their escape routes will be treacherous and their fates loathsome once apprehended can counterterrorism systems be effective deterrents.[7] For those governments sponsoring terrorism, the responsible officials should be on notice that their participation brands them as terrorists and catapults them into the crosshairs of punishment—painful, quick punishment by the international community, in sharp constrast to the delays and procrastinations of the past.

Perhaps our covert services need to be more ingenious in disrupting terrorist preparations with methods that are not quite so far out of bounds as assassinations. Experts have suggested spraying terrorists with immobilizing foam, drugging them with sleep-inducing mists, bombarding their hideaways with low-frequency noise that disorients them, or blinding them with lasers. Others have proposed that their equipment be coated with caustic chemicals, or sent power surges that knock out their sensitive electronic equipment. Such actions surely seem like the grist for Hollywood phantasmagoria, but they have been put on the table for consideration.[8]

That said, more than a dozen committees of the U.S. Congress are addressing the dangers on the horizon. They now recognize that terrorism has irreversibly moved from a distant international issue to an immediate domestic threat, with the FBI and local law enforcement agencies instructed to be better prepared with shorter response times. They emphasize protection of people and property in the United States, with airport security and safety in federal buildings being special priorities. They call for our military forces to throw some of their weight behind countering terrorism, and they fund our scientists' research with large sums of money to help them find smart technologies, which will thwart new enemies targeted on America.[9]

Whatever the response, in the near term terrorists will not only persist but will wreak even greater damage as the destructive power of their arsenals increase. We have no choice but to meet force with force and place more barriers in their paths. As we do, will terrorism simply become "part of the cost of doing business in *free* societies?" Need we change our concept of freedom? We must honestly ask ourselves some piercing questions:

- How free are we in a society where architects are reluctant to design buildings with underground parking garages because of their vulnerability to bomb attacks?
- Is it really a free society that wraps layers of security blankets around its own national celebrations, let alone international competitions, for fear of terrorist bombings?
- Are we really free when we question every unattended piece of luggage on a train or in an airport, and we spend more and more time being tracked on cameras deployed by security personnel?
- What kind of freedom surrounds our children when they are the innocent prey of drug lords and their criminal associates?

As the nation's "paranoia pulse" rises, we have no choice but to accept a redefinition of personal freedom, not only when entering airports but in other areas of our daily lives as well.

During the very week I put the finishing touches on this book, a spate of terrorist incidents around the world further reminded me of the dangers surrounding us. Opponents of Georgian President Eduard Shevardnadze seized four UN observers as hostages to force concessions from the government. Unknown bombers destroyed a General Motors dealership in Greece, protesting U.S. policy toward Iraq. Dissidents carried out an attack on a religious shrine in Sri Lanka. A train bombing in Algeria and a street bombing in Ulster rounded out a typical seven days of what has been called "primitive terrorism."[10]

As to superterrorism, the arrest in Las Vegas of a white supremacist believed to be in possession of anthrax spores (which turned out to be harmless anthrax vaccine) stole the headlines. But a more worrisome development was related to a report of four hundred shipments of botulinum toxin from a California pharmaceutical company to uncertain destinations. On the international scene, the Russian government was accused of shipping additional nuclear equipment to Iran with possible military applications; Russian researchers were called to task for developing a strain of anthrax bacteria that could overwhelm any known vaccine, the implication being that the strain was to be used for hostile purposes.[11]

On television, *Prime Time* devoted one hour to the dangers of germ warfare—recounting scenarios of potential devastation originating in Siberia, Baghdad, and Las Vegas.[12] In addition, every news broadcast was full of the dangerous adventurism of Saddam Hussein, the military brinksmanship of the United States, and the mediation efforts of the UN Secretary General.[13]

Of course, at this point in history, no book about terrorism would be complete without a commentary on Iraq. The situation is so fluid that any remarks about the immediate crisis would soon be out of date. However, two longer-term aspects of the situation seem clear. First, no matter how many times UN inspectors or U.S. aircraft destroy Iraq's chemical and biological weapons stockpiles and facilities, assuming they find them, the threat of biological and chemical weapons will persist; many highly trained specialists in the country will simply rebuild—as often as necessary and whenever they can get away with it. Thus, it is imperative to try to redirect the efforts of the scientists and engineers themselves to other pursuits so they do not have the time or interest for the rebuilding game. Second, Israel will remain the focus of Iraqi hatred for the indefinite future. Unless this hatred is tempered, the motivation for terrorism will remain.

It has been suggested that to reduce the likelihood of superterrorism springing from Iraq, or indeed Iran or other neighbors, the Middle East should become a zone that is free of biological and chemical weapons.[14] One possible method for accomplishing this would see the countries of the region, including Israel, renouncing the use of all such weapons in accordance with the chemical and biological weapons conventions. By doing so, these nations would accept the principle of international transparency of facilities that have the capability of contributing to such weaponry, and they would embrace a program of redirection of military scientists to peaceful endeavors. If this approach were successful, in time regional arrangements to address nuclear weapons might also be considered, recognizing the difficulty in persuading Israel to give up its nuclear arsenal.

These steps toward freeing the region of weapons of mass destruction and weaning the specialists away from weapons development and into cooperative programs of more positive benefit may seem far away and far from perfect. But they would go a long way to deterring weapons buildups, while serving as important complements to broader policies and other programs that will undoubtedly be adopted.

In sum, unless the projected trajectory of terrorism is somehow capped, we doom ourselves to a future increasingly punctuated with assaults on our physical safety, regardless of the security barricades. Indeed, the only end-game that makes sense is one that redirects the momentum of terrorism toward building rather than destroying the nations of the world—an end-game that will take decades to accomplish. While there are many champions in Washington for individual items on the counterterrorism agenda, our political leaders must apply more creative thinking to the phenomena underlying many terrorist acts that will increasingly wash up on our shores: the geographic displacement of ethnic and religious groups, the ever-growing gap between rich and poor, the spread of destructive technologies to all corners of the globe, the growth of international organized crime entrenched in hospitable havens around the world, and the growing acceptance of corruption as an inevitable trait of modern societies.

I underscore that in order to reduce the seeds from which terrorism sprouts, promoting democracy and open economics is critical. Free enterprise and job creation are key ingredients that can be designed to attract young people who are not already completely brainwashed by the hatred fueling other generations. Also, for the hundreds of millions of deprived people in the poor countries of the world, access to their own land and water is a second essential. Indeed, they will not be denied in their quest for greater economic parity with more prosperous populations and for greater respect for their ethnic and religious heritages. They demand realignment of borders and resettlement rights in distant lands, including the United States. A foreign policy to accommodate such aspirations has yet to be developed or even discussed.

Even if such policies were to be effective, there will always remain those factions and individuals who are not satisfied with their piece of the action. They seek dominance, rather than equality, and power, rather than participation. No foreign policy can consider accommodation when such unfulfilled expectations stem not from realistic needs but rather from radical ideology or from religious or nationalistic motivations based on prejudice and hatred. The behavior inevitably arising out of such misguided thinking will continue to be our challenge for the future. It must be identified for what it is, be it crime or terrorism, and dealt with accordingly.

Nevertheless, we have no choice but to invest resources in raising the *legitimate* expectations of deprived populations that are increasingly turning to violence as their only route of escape from lives of subjugation, misery, and unfulfilled expectations. Economic progress will not drown out the ethnic and religious hatred that fuels much of the terrorism around the world, but it can provide new incentives to start the process of political accommodation. With superterrorism rapidly becoming a reality, the alternative to seeking new paths to peace is for the future of all societies to be doomed by internecine warfare of the worst kind.

Endnotes

PROLOGUE

1. Field notes of the participants.
2. *Patterns of Global Terrorism,* United States Department of State, April 1997. Also, comments by a senior FBI official during discussions with the author on July 24, 1997, Washington, D.C.
3. Comments by senior FBI official. See also, Robert Suro, "2 Terrorist Groups Set Up U.S. Cells, Senate Panel Is Told. 2,600 FBI Positions Dedicated to the Fight, FBI's Freeh Says," *The Washington Post,* May 14, 1997, p. A4.
4. Dr. Steven J. Hatfill, National Institutes of Health, during his illustrated presentation at a public seminar "Terrorism and Civil Defense" at George Washington University, August 12, 1997.
5. Professor Gideon Frieder, Department of Engineering, George Washington University, during his presentation at a public seminar "Terrorism and Civil Defense" at George Washington University, August 12, 1997. Also, for discussion of the Bulgarian virus see David S. Bennahum, "Heart of Darkness," *Wired,* November 1997, p. 226.
6. Fox Butterfield, "New Devices May Let Police Spot People on the Street Hiding Guns," *The New York Times,* April 8, 1997, p. 1.
7. See Mark Hansen, "No Place To Hide," *Journal of the American Bar Association,* August 1997. Using information from the National Institute of Justice, this article reports that within five years new scanning devices should enable the police to conduct the equivalent of a strip search at a distance of sixty feet. A camera employing passive millimeter wave imaging technology should be able to see through clothes and building material. According to predictions a radar skin scanner will produce anatomically correct, detailed images of the body.
8. *Global Proliferation of Weapons of Mass Destruction,* Hearings before the Permanent Subcommittee on Investigations of the Committee on Governmental Affairs, United States Senate, Part I, October 31 and November 1, 1995, Staff Report, p. 67. "The Occurrence of Sarin Gas Murders and Other Extraordinary Incidents, and the Police Response," *White Paper on Police 1995 (Excerpt),* National Police Agency, Government of Japan, Undated, obtained from National Police Agency in 1997.
9. Nicholas D. Kristof, "Japanese Said to Have Planned Nerve-Gas Attacks in U.S.," *The New York Times,* March 23, 1997, p. 1.

10. *Global Proliferation of Weapons of Mass Destruction*, Hearings before the Permanent Subcommittee on Investigations of the Committee on Governmental Affairs, United States Senate, Part III, March 27, 1996, Sentencing Statement of Judge Duffy, pp. 276–277.

11. William Laqueur, *The Age of Terrorism*, Little-Brown, Boston, 1987, p. 313.

12. Comments by Glenn C. Schoen, Director of Analytical Services for International Security Management, Inc., in his class, "Understanding Contemporary Terrorism," Georgetown University, Spring Semester, 1997.

13. "Vast Negligence Reported in Granting of Citizenship," *The New York Times*, February 25, 1997, p. 1. According to this story, 180,000 immigrants were granted citizenship without required background checks. Of those checked, 72,000 had criminal records, and 10,000 of them had been arrested for felonies.

14. Discussion with Department of State senior official responsible for visa issuance policy, July, 1997.

15. Announcement by FBI Director Louis Freeh, C-Span television broadcast, April 19, 1997.

16. Ibid.

17. *Global Organized Crime*, Center for Strategic and International Studies, Washington, D.C., 1994, Appendix.

18. Many specialists cling to the belief there should be a sharp distinction between organized crime that is intent on obtaining wealth and terrorist organizations that are focused on political retribution. For a good discussion of the increasing convergence of these two types of activities, see the following document prepared in connection with the World Ministerial Conference on organized crime held in Naples, Italy, November 1994: "Problems and Dangers Posed by Organized Transnational Crime in the Various Regions of the World," *Transnational Organized Crime*, Autumn 1995, pp. 24–26. Also, see Chapter 10 of this book.

CHAPTER 1
MODERN MUTATIONS OF TERRORISM

1. Many books have been written about the history of terrorism. For example, an excellent summary of the principal events is contained in Jeffrey D. Simon, *The Terrorist Trap*, Indiana University Press, Bloomington and Indianapolis, 1994, pp. 13–166. A more comprehensive treatment of the topic is contained in Martha Crenshaw and John Pimlott (editors), *Encyclopedia of World Terrorism*, M. E. Sharpe, Armonk, NY, 1997, Volumes 1, 2, and 3. I have not attempted to summarize such writings, but I have presented a very limited number of events which help illustrate the deeply embedded roots and the many facets of terrorism as well as point out current trends that are reorienting thinking and attitudes about the phenomenon that is frequently referred to as "political violence."

2. *Patterns of Global Terrorism*, Department of State, April 1997, p. vi.

3. Paul Johnson, "The Seven Deadly Sins of Terrorism," included in Henry H. Hahn (editor), *Terrorism and Political Violence: Limits and Possibilities of Legal Control*, Oceana Publications, New York, 1990, pp. 189–190.

4. Ibid.

5. David E. Long, *The Anatomy of Terrorism*, The Free Press, New York, 1990, p. 2.

6. Simon, p. 26.

7. R.R. Palmer, *Twelve Who Ruled: The Commission of Public Safety during the Terror*, Princeton University Press, Princeton, 1941, p. 226.

8. George Rude (editor), *Robespierre*, Prentice Hall, New York, 1967, p. 8.

9. Palmer, p. 260.

10. Simon, p. 29.

11. Ibid, p. 32.

12. Ibid., p. 40.

13. Walter Laqueur, *The Age of Terrorism*, Little-Brown, Boston, 1987, p. 19.

14. Walter Laqueur, "Post Modern Terrorism," *Foreign Affairs*, September–October 1996, p. 24.

15. Johnson, pp. 190–191.

16. Comments by Glenn C. Schoen, Director of Analytical Services for International Security Management, Inc., in his class, "Understanding Contemporary Terrorism," Georgetown University, Spring 1997.

17. Roger Medd, "International Terrorism on the Eve of the New Millennium," *Studies in Conflict and Terrorism*, Vol. 20, pp. 280–282, No. 3, July–September, 1997.

18. Twelve years later, the PLO agreed to a settlement of claims with the family of Leon Klughoffer and the Crown Travel Services, Inc., "PLO Settles with Family of Achilee Lauro Victim," Associated Press, *The Washington Post*, August 12, 1997, p. A9.

19. Simon, p. 107.

20. Laqueur, *The Age of Terrorism*, p. 7–8.

21. John Kerry, *The New War: The Web of Crime that Threatens America's Security*, Simon and Schuster, New York, 1997, p. 176.

22. Laqueur, *The Age of Terrorism*, p. 9.

23. Ibid, p. 12.

24. Bruce Hoffman, "Responding to Terrorism Across the Technological Spectrum," Strategic Studies Institute, U.S. Army War College, Carlisle Barracks, Pennsylvania, July 15, 1994, pp. 15–21.

25. Stephen Sloan, "Terrorism: How Vulnerable Is the United States," in Stephen Pelletiere (editor), *Terrorism: National Security Policy and the Home Front*, Strategic Studies Institute, U.S. Army War College, Carlisle Barracks, Pennsylvania, May 1995, p. 35

26. Medd, p. 282.

27. Robert J. Kelly and Rufus Schatzberg, "Galvanizing Indiscriminate Political Violence: Mind-sets and Some Ideological Constructs in Terrorism," *International Journal of Comparative and Applied Criminal Justice*, Spring 1992, Vol. 16, No. 1, p. 13.

28. Jeffrey Ian Ross, "The Psychological Causes of Oppositional Political Terrorism: Toward an Integration of Findings," *International Journal of Group Tensions*, 1994, Vol. 24, No. 2, p. 160.

29. Maxwell Taylor and Ethel Quayle, *Terrorist Lives*, Brassey's, Washington-London, 1994, p. 57.

30. Ibid.

31. Donna M. Schlagheck, in a book review, discusses the debate over the uniqueness of each terrorist group, in *Studies in Conflict and Terrorism*, Vol. 20, No. 1, 1997, p. 126.

32. Long, p. 3.

33. "Excerpt from Statements," *The New York Times,* January 6, 1998, p. A 15.

34. Laqueur, "Postmodern Terrorism," p. 24.

35. CNN Broadcast from Lima, Peru, April 25, 1997.

36. See Raphael F. Perl, "Terrorism, the Media, and the Government: Perspectives, Trends, and Options for Policymakers," *CRS Report for Congress* #97-960F, October 22, 1997.

37. For a discussion of Russian arms exports, see Andrew J. Pierre and Dmitri V. Trevin, (editors), *Russia, In the World Arms Trade,* Carnegie Endowment for International Peace, Washington, D.C., 1997.

38. *Russian Organized Crime,* Center for Strategic and International Studies, Washington, D.C., September 1997.

39. Raphael F. Perl, "Terrorism, the Future, and U.S. Foreign Policy," *CRS Issue Brief,* IB95112, November 24, 1997.

40. See, for example, *Proliferation: Threat and Response,* Section on Transnational Threats, Department of Defense, November 1997.

41. Phil Williams, "Getting Rich and Getting Even, Transnational Threats in the Twenty-First Century," background paper prepared in connection with the report Linda P. Raine and Frank J. Cilluffo (editors), *Global Organize Crime: The New Empire of Evil,* Center for Strategic and International Studies, Washington, D.C., 1994.

42. Medd, p. 291.

43. Linda P. Raine and Frank J. Cilluffo (editors), *Global Organized Crime*, Center for Strategic and International Studies, Washington, D.C., 1994.

44. Author's interview with Department of State counterterrorism official, April 1997.

45. Ibid.

46. Richard Clutterbuck, *Terrorism in an Unstable World,* Routledge, London, and New York, 1994.

47. Kerry, p. 27.

48. Author's interview with Finance Minister of Indonesia, April 1989.

49. Williams.

50. Author's interview with senior Department of Defense official, July 1997.

51. Medd, p. 289.

52. Author's interview with FBI spokesperson, February 1997.

53. Philip C. Wilcox, "International Terrorism," remarks before the Denver Council on Foreign Relations, September 12, 1996, released by the Department of State, September 1996. See also Sloan, p. 1.

54. A provocative discussion of the boundaries of terrorism was released on the home page of *The Economist* in February 1996. It persuasively argued that the act or the person that deserves the label of terrorism or terrorist depends on who is bestowing the label. For example, Sinn Fein considered all aspects of IRA violence to be legitimate warfare while most Arabs consider the Israel Defense Force to be the worst terrorists in the Middle East. Of course it is difficult for Americans to accept the proposition that the worst single terrorist acts of all time were the nuclear bombings of Nagasaki and Hiroshima. Identifying entire governments as terrorist organizations is a growing trend. At the same time, it may be much easier to disarm a rogue government with relatively well-defined responsibilities than a subnational terrorist group which is more difficult to identify.

55. John Deutch, "Terrorism," *Foreign Policy,* Fall 1997, p. 21.

CHAPTER 2
THE NUCLEAR LEGACY: A SHOPPING MALL FOR ROGUE STATES AND TERRORISTS

1. Oleg Bukharin and William Potter, "Potatoes Were Guarded Better," *The Bulletin of the Atomic Scientists,* May/June 1995, p. 46–50; and Richard and Joyce Wolkomir, "Where Armageddon Is All in a Day's Work," *Smithsonian,* January–February 1997, pp. 115–127. Other Russian and American publications have also reported this incident; while minor details of the event are reported differently, the main thrust of the story is the same in all reports.
2. Bukharin and Potter, p. 48.
3. David Hoffman, "Banditry Threatens the New Russia," *The Washington Post,* May 12, 1997, p. Al.
4. *Global Proliferation of Weapons of Mass Destruction,* Hearings, Permanent Sub-Committee on Investigations, Committee on Governmental Affairs, U.S. Senate, Part II, March 13, 20, and 22, 1996, p. 27.
5. For background material related to *The Peacemaker,* see Andrew and Leslie Cockburn, *One Point Safe: A True Story,* Anchor Books, New York, 1997. For allegations by Alexander Lebed, see Laura Myers, "Yeltsin Foe Says Russia Lost Nukes," *Associated Press,* September 5, 1997.
6. William M. Arkin, "No Points Safe," *The Bulletin of the Atomic Scientists,* January–February, 1998, p. 73.
7. Statement by Captain Mike Doubleday, Daily News Briefing, the Pentagon, October 2, 1997.
8. Walter Pincus, "U.S. Developed 60-Pound Nuclear Weapon a Parachute Could Deploy," *The Washington Post,* December 23, 1997, p. A4. Also, *NBC Nightly News with Tom Brokaw,* October 7, 1997; this report was consistent with official statements by spokespersons from the CIA and Department of Defense.
9. John Deutch, "The Threat of Nuclear Diversion," *Global Proliferation of Weapons of Mass Destruction,* Hearings, Committee on Governmental Affairs, United States Senate, Part II, March 13, 20, and 22, 1996, pp. 302–323.
10. Ibid.
11. "Rodionov: Russia Is Losing Control of Nuclear Forces," OMRI electronic mail service reporting developments in Russia, February 7, 1997.
12. For additional details on the plight of soldiers and the spread of corruption in the military forces, see as examples of the many analyses of the problem two reports by Graham Turboville, "Organized Crime and the Russian Armed Forces," *Transnational Organized Crime,* vol. 1, no. 4, Winter 1995, pp. 73–77, and "Weapons Proliferation and Organized Crime: The Russian Military and Security Force Dimension," USAF Institute for National Security Studies, Proliferation Series, Occasional Paper #10, June 1996.
13. Martin Nesirky, "U.S. General Says Russia To Shadow Nuclear Staff," *Reuters News Service,* Russia, October 28, 1997.
14. "The Soviet Nuclear Threat Reduction Act of 1991," commonly referred to as the Nunn–Lugar Initiative. Also see John Shields and William Potter (editors), *Dismantling the Cold War,* MIT Press, Cambridge, Massachusetts, 1997.
15. See *Proliferation Concerns: Assessing U.S. Efforts to Help Control Nuclear and Other Dangerous Material and Technologies in the Former Soviet Union,* National Research Council, National Academy Press, Washington 1997, Chapter 3.

16. Highly enriched uranium, or *HEU*, refers to unirradiated uranium enriched to a level of uranium[235] of at least 20 percent. Plutonium refers to separated plutonium. These materials can be used in nuclear weapons without the need for complicated chemical processing. Uranium enriched to higher levels and plutonium produced explicitly for weapons purposes will of course be most effective. As noted, a minimum of several kilograms of plutonium and several times that amount of *HEU* are required for a nuclear weapon using state-of-the-art design and construction techniques. The actual amount will vary significantly depending on the exact composition of the material, the type of weapon, and the sophistication of the design. In the past, crude weapons have usually used many times the minimum amount cited above. For a discussion of the technical aspects of designing and building a nuclear weapon, see Graham T. Allison, Owen R. Cote, Jr., Richard A. Falkenrath, and Steven E. Miller, *Avoiding Nuclear Anarchy*, MIT Press, Cambridge, 1995, Appendix B.

17. Deutch, p. 310.

18. See for example, *op. cit., Proliferation Concerns*, Chapter 2.

19. Observations during visit to Russian nuclear facilities where upgrading of security was in progress, May 1996.

20. Ibid.

21. Ibid.

22. *Proliferation Concerns*, Chapter 2. Also, *MPC&A Strategic Plan*, Department of Energy, January 1998.

23. Ibid.

24. Shields and Potter, p. 345. For a more popularized account, see Brian Eads, "A Shopping Mall for Nuclear Blackmailers", *The Readers Digest*, April, 1997, pp. 173–184. Also for the official version, see *Project Sapphire: Transferring Vulnerable Nuclear Materials to Safe Storage*, U.S. Department of Energy, March 1998.

25. Author's interviews in Almaty and Moscow, April 1996.

26. Author's interviews in Tbilisi, Georgia, October 1996.

27. Visit to Institute of Nuclear Physics, Tbilisi, Georgia, October 1996.

28. Michael R. Gordon, "Russia Thwarting U.S. Bid To Secure a Nuclear Cache," *The New York Times*, January 5, 1997, p. A16. "U.S. Helps To Remove Georgian Uranium," *The Washington Post* (Provided by Reuters), April 22, 1998, p. A14.

29. Author's interview in Moscow with Russian export control expert, November 1996.

30. The number of whistleblowers in Russia is on the rise, with many being sufficiently bold to protest previously secret activities through "illegal" whistleblowing. A well advertised, and "legal" reward system should be quite effective in encouraging whistleblowing.

31. *Proliferation Concerns*, Chapter 2.

32. See, for example, *Proliferation of Weapons of Mass Destruction, Assessing the Risks*, Office of Technology Assessment, United States Congress, OTA-ISC-559, 1993.

33. See "The Nuclear Black Market, Global Organized Crime Project," Center for Strategic and International Studies, Washington, 1996.

34. *Proliferation: Threat and Response*, Department of Defense, November 1997, and *The Proliferation Primer*, a majority report of the Subcommittee on International Security, Proliferation, and Federal Services, Committee on Governmental Affairs, U.S. Senate, January 1998.

35. Briefing by American experts at Sandia National Laboratory, Albuquerque, New Mexico, April 1996.

36. Ibid.
37. Ibid.
38. "Testimony of Ambassador Rolf Ekeus, *"Global Proliferation of Weapons of Mass Destruction,"* Hearings, Committee on Governmental Affairs, Part II, March 13, 20, and 22, pp. 90–105.
39. David Kay, "Iraq and Beyond: Understanding the Threat of Weapons Proliferation," *Global Proliferation of Weapons of Mass Destruction,* Hearings, Committee on Governmental Affairs, United States Senate, Part II, March 13, 20, and 22, 1996, pp. 324–332.
40. "U.N. Agency Head Sees No Iraqi Nuclear Activity," *The Washington Post,* December 4, 1997, p. A34. U.S. caution due to lack of access to all Iraqi documents is reflected in *Inspections in Iraq: Preventing Iraqi Reconstitution of Its Nuclear Weapons Program,* U.S. Deparatment of Energy, March 1998.
41. *Global Proliferation of Weapons of Mass Destruction,* Hearings, p. 935.
42. Alexander Y. Masourov, "The Origins, Evolution, and Current Politics of the North Korean Nuclear Program," *The Nonproliferation Review,* vol. 2, no. 3, Spring/Summer 1995, Monterey Institute of International Studies, pp. 25–38.
43. Ibid.
44. Ibid.
45. For additional background, see David Albright, "North Korea and the 'Worst-Case Scare-nario,'" *The Bulletin of the Atomic Scientists,* January/February 1994, pp. 3–7. An official U.S. view on developments was presented in a briefing by Ambassador Robert Gallucci, Carnegie Endowment for International Peace, Washington, April 1995. For a comprehensive overview of recent negotiations, see Leon D. Sigal, *Disarming Strangers, Nuclear Diplomacy with North Korea,* Princeton University Press, Princeton, 1997.
46. Ibid.
47. *Proliferation of Weapons of Mass Destruction, Assessing the Risks,* Office of Technology Assessment, United States Congress, Government Printing Office, August 1993, p. 64.
48. Since the Indian and Pakistani testing occurred just as this manuscript was being put into final form for publication, only very preliminary reactions were available. The principal sources for the brief commentary were American newspaper and television accounts that were manifold at the time of the Indian and Pakistani announcements. As one example, see Teresita C. Schaffer and Howard B. Schaffer, "After India's Tests, and Now Pakistan's," *The Washington Post,* May 29, 1998, p. A27. For earlier discussions that seemed to anticipate developments in South Asia, see "Carnegie Quarterly," Carnegie Corporation of New York, Vol. XLI/No. 2–3, Spring–Summer, 1996, and Richard N. Haas and Gideon Rose, "Facing the Nuclear Facts in India and Pakistan," *The Washington Post,* Outlook Section, January 6, 1997, p. 2. Also many related issues were aired in author's interviews during Indian–Pakistani conference in Bellagio, Italy, June 1996.
49. David Albright, "South Africa and the Affordable Bomb," *The Bulletin of the Atomic Scientists,* July/August 1994, pp. 37–47.
50. Author's visit to Vietnam in April 1972.
51. R. Jeffrey Smith, "U.S. Details Gaffe in Retrieving Nuclear Material from Vietnam," *The Washington Post,* January 16, 1997, p. A22.
52. For additional background on NEST, see "Prepared Statement of Duane C. Sewell," *Global Proliferation of Weapons of Mass Destruction,* Hearings, Commit-

tee on Governmental Affairs, United States Senate, Part III, March 27, 1996, pp. 70–72.

53. "Opening Statement of Senator Lugar," *Global Proliferation of Weapons of Mass Destruction,* Hearings, Committee on Governmental Affairs, United States Senate, Part II, March 13, 20, and 22, 1966, p. 11.

54. With regard to the Islamic Jihad, see *The Nuclear Black Market,* Center for Strategic and International Studies, Washington, p. 15. With regard to the PLO, this rumor has been circulating in Washington for several years, but I am unaware of any evidence confirming it.

55. *Global Proliferation of Weapons of Mass Destruction,* Hearings, Permanent Subcommittee on Investigations, Senate Committee on Governmental Affairs, Part I. October 31, November 1, 1995, p. 72–73.

56. There are many interpretations of events at Reykjavik. See particularly, Henry A. Kissinger, *Diplomacy,* Simon and Schuster, New York, 1994, p. 783.

57. For the generals' views, see Craig Cerniello, "Retired Generals Re-ignite Debate over Abolition of Nuclear Weapons," *Arms Control Today,* November/December 1996, pp. 14–18.

58. See, for example, "Don't Ban the Bomb," *The Economist,* January 4, 1996, p. 15–16.

CHAPTER 3
CHEMOTERRORISM: POISONING THE AIR, WATER, AND FOOD SUPPLY

1. John Cloud, "Margin of Terror," *City Paper,* Washington, D.C., March 14–20, 1997, Vol. 17. No. 10, pp. 18–32.

2. Ibid.

3. Nicholas Kristof, "Japanese Cult Said to Have Planned Nerve-Gas Attacks in U.S.," *The New York Times,* March 23, 1997, p. 1.

4. David Hanson, "Simulated Chemical Attack on Nation's Capitol," *Chemical and Engineering News,* May 5, 1997, p. 12.

5. "Staff Statement, Global Proliferation of Weapons of Mass Destruction: A Case Study on the Aum Shinrikyo," *Global Proliferation of Weapons of Mass Destruction,* Hearings, Permanent Subcommittee on Investigations, Committee on Governmental Affairs, United States Senate, Part I, October 31 and November 1, 1995, pp. 47–102.

6. See, for example, David Kaplan and Andrew Marshall, *The Cult at the End of the World,* Crown, New York 1996; and D.W. Brackett, *Holy Terror: Armageddon in Tokyo,* Weatherill, New York 1996.

7. Staff Statement, p. 47.

8. John Deutch, *Global Proliferation of Weapons of Mass Destruction,* Hearings, Permanent Subcommittee on Governmental Affairs, United States Senate, Part II, March 13, 20, 22, 1996, p. 75.

9. Staff Statement, p. 49.

10. "Prepared Statement of Mr. Kyle B. Olson," *Global Proliferation of Weapons of Mass Destruction*, Hearings, Permanent Subcommittee on Investigations, Committee on Governmental Affairs, United States Senate, Part I, October 31 and November 1, 1995, pp. 109–112.
11. Staff Statement, p. 53.
12. Ibid, p. 72.
13. "White Paper on Police 1995 (Excerpt)," National Police Agency, Government of Japan, Translated and Published by Police Association, Tokyo, Distributed by Japanese Embassy in Washington, September 1997. Also, Kevin Sullivan, "Aum Shinrikyo Rebuilds Its Following in Japan," *International Herald Tribune*, September 30, 1997, p. 2.
14. Leonid Krutakov and Ivan Kadulin, "Seko Asahara's Russian Trial," *Passport*, Moscow, No. 5, 1997, p. 28.
15. Ron Purver, "The Threat of Chemical and Biological Terrorism," *The Monitor*, University of Georgia, Spring, 1997, Vol. 3, No. 2, p. 7.
16. "Statement of Robert M. Blitzer," Hearing on Chemical-Biological Defense Program and Response to Urban Terrorism, National Security Committee, Subcommittee on Military Research and Development, U.S. House of Representatives, March 12, 1995, available from the FBI.
17. Joby Warrick, "A Toxic Bargain's Continuing Costs," *The Washington Post*, August 18, 1997, p. A1.
18. *The 1997 GenCon Conference on Improving U.S. Capabilities for Defense from Bioterrorism, January 16, 1997*, The National Consortium for Genomic Resources Management and Services, Montros, Virginia, February 1997, Presentation by Jerome Hauer, Director, New York City Office of Emergency Services, p. 27.
19. Jim Hoagland, "Poison Gas for an Israeli Assassination. What Could They Have Been Thinking?" *International Herald Tribune*, October 10, 1997, p. 9; Barton Gellman, "Botched Assassination by Israel Gives New Life to Hamas," *The Washington Post*, October 6, 1997, p. A1.
20. Author's interview with Libyan exchange visitor, Moscow, Fall 1992.
21. See, for example, *The Chemical and Biological Warfare Threat*, Central Intelligence Agency, released in 1995; *Medical Management of Chemical Casualties*, Chemical Casualty Care Office, Medical Research Institute of Chemical Defense, Aberdeen Proving Ground, Maryland, September 1995. Also, "Prepared Statement of James A. Genovese," *Global Proliferation of Weapons of Mass Destruction*, Hearings, Permanent Subcommittee on Investigations, Committee on Governmental Affairs, United States Senate, Part I, October 31 and November 1, 1995, pp. 122–126.
22. Ibid, Genovese, p. 125.
23. Ibid, p. 123.
24. *Medical Management of Chemical Casualties*, p. 4.
25. Ibid, p. 8.
26. *The Chemical and Biological Warfare Threat*. For a discussion of K agents, see Leonard A. Cole, *The Eleventh Plague*, W.H. Freeman, New York, 1996, p. 31.
27. *Medical Management of Chemical Casualties*, pp. 6–7.
28. Author interviews with members of Russian Duma responsible for providing funds for destruction of chemical weapons, Moscow, Fall 1996.

29. Amy E. Smithson, Dr. Vil S. Mirzayanov, Maj. Gen. Roland Lajoie (USA, Ret.), Michael Krepon, *Chemical Weapons, Disarmament in Russia: Problems and Prospects,* The Henry L. Stimson Center, Report No. 17, October 1995.

30. Author's interview with Russian Duma expert, Fall 1996.

31. "Report of the Secretary General on the Status of the Implementation of the Special Commission's Plan for the Ongoing Monitoring and Verification of Iraq's Compliance with Relevant Parts of Section C of Security Resolution 687 (1991)," S/1995/864, United Nations, New York, pp. 16–19.

32. Ibid.

33. For extended discussion see Cole, pp. 81–84.

34. *The 1997 GenCon Conference,* Presentation by David Kay, p. 35.

35. "Report of the Secretary General," pp. 16–19.

36. "Czechs Told U.S. They Detected Nerve Gas During the Gulf War," *The New York Times,* October 19, 1996, p. 1.

37. William J. Perry, and Janet Reno, "A Treaty in the U.S. Interest," *The Washington Post,* September 11, 1996, p. A23.

38. "The Chemical Weapons Convention," Fact Sheet, U. S. Arms Control and Disarmament Agency, December 17, 1996.

39. "Chemical Weapons Treaty Blocked," *Chemical and Engineering News,* September 16, 1996, p. 6. Also, Lally Weymouth, "Chemical Weapons Fraud," *The Washington Post,* September 12, 1996, p. A20.

40. See, for example, Charles Krauthammer, "A False Sense of Security," *The Washington Post,* April 11, 1997, OpEd page.

41. Speech by Secretary Madeline Albright at James Baker Center, Rice University, Houston, February 1997, available from Department of State.

42. See, for example, Michael Heylin, "Chemical Arms Pact Ratified," *Chemical and Engineering News,* April 28, 1997, p. 9; and Helen Dewar, "Senate Approves Chemical Arms Pact after Clinton Pledge," *The Washington Post,* April 25, 1997, p. A1.

43. "BXA Facts, The Australia Group," Fact Sheet, Bureau of Export Administration, Department of Commerce, 1996.

44. For a comprehensive discussion of U.S. efforts to regulate toxic substances, see Glenn E. Schweitzer, *Borrowed Earth, Borrowed Time, Healing America's Chemical Wounds,* Plenum, New York, 1991.

45. Ibid, pp. 206–207.

46. "FBI Takes Lead in Developing Counterterrorism Effort," *Chemical and Engineering News,* November 4, 1996, pp. 10–14. See also, *The 1997 GenCon Conference,* Presentation by Randall S. Murch, FBI Scientific Laboratory, p. 5.

47. Ibid.

48. *Medical Management of Chemical Casualties*

49. Garry L. Briese, Letter to Federal Emergency Management Agency, January 31, 1995, as cited in *Global Proliferation of Weapons of Mass Destruction,* Hearings, Permanent Subcommittee on Investigations, Committee on Governmental Affairs, United States, Senate, Part III, March 27, 1996, p. 224–5.32.

50. *The 1997 GenCon Conference,* Presentation by Kenneth Meyers, Office of Senator Richard Lugar, p. 37.

51. Cole, p. 157.

52. *The 1997 GenCon Conference,* Presentation by Jerome Hauer, p. 28.

CHAPTER 4
BIOTERRORISM: THE LAST FRONTIER

1. See Sari Horwitz, "B'nai B'rith Package Contained Common Bacteria," *The Washington Post*, April 29, 1997, p. B2; "A Leaking Package at B'nai B'rith Makes People Sick," posted on World Wide Web by CNN, April 24, 1997; "FBI: 'Leaking Package at B'nai B'rith Not Life Threatening'," posted on World Wide Web by CNN, April 24, 1997.
2. Ibid.
3. Ibid.
4. Author's interviews with leading Russian ecologist, Yekaterinburg, Russia, November 1992 and June 1994.
5. Matthew Meselson, Jeanne Guillemin, Martin Hugh-Jones, Alexander Langmuir, Ilona Popova, Alexis Shelokov, Olga Yampolskaya, "The Sverdlovsk Anthrax Outbreak of 1979," *Science*, November 18, 1994, pp. 1202–1208. The following description of anthrax is presented in *Medical Management of Biological Casualties*, Handbook, U.S. Army Medical Research Institute of Infectious Diseases, Fort Detrick, Maryland, March 1996, p. 17:

   > *Bacillus anthracis* is a rod-shaped, gram-positive, sporulating organism, the spores constituting the usual infective form. Anthrax is a zoonotic disease with cattle, sheep, and horses being the chief animal hosts but other animals may be infected. The disease may be contracted by the handling of contaminated hair, wool, hides, flesh, blood and excreta of infected animals and from manufactured products such as bone meal, as well as by purposeful dissemination of spores. Transmission is made through scratches or abrasions of the skin, wounds, inhalation of spores, eating insufficiently cooked infected meat, or by flies. All human populations are susceptible. Recovery from an attack of the disease may be followed by immunity. The spores are very stable and may remain viable for many years in soil and water. They will resist sunlight for varying periods.

6. For Yeltsin admission, see D. Muratov, Y. Sorokin, and V. Fronin, "Boris Yeltsin: I Am Not Hiding the Difficulties and I Want People to Understand Them," *Komsomolskaya Pravda*, May 27, 1992, p. 2. In Article I, the *Convention on the Prohibition of the Development, Production, and Stockpiling of Bacteriological (Biological) and Toxin Weapons and on Their Destruction* states:

   > "Each State Party to this Convention undertakes never in any circumstances to develop produce, stockpile, or otherwise acquire or retain: (1) Microbial or other biological agents or toxins whatever their origin or method of production, of types and in quantities that have no justification for prophylactic, protective or other peaceful purposes . . ."

7. Author's interview with senior U.S. official responsible for U.S. policy concerning biological research activities in Russia, Washington, D.C., March, 1997.

8. Meselson. Also, N. Wade, "New Research in Soviet Anthrax Leads to Questions about Vaccine," *The New York Times*, February 3, 1998, tends to confirm the laboratory origin of the anthrax.

9. Ibid. With regard to vaccinations, see *Medical Management of Biological Casualties*, p. 16.

10. Author's interviews with American specialists in weaponization of biological agents, Washington, D.C., April and May, 1997.

11. Testimony of Ambassador Rolf Ekeus, Executive Chairman, United Nations Special Commission, *Global Proliferation of Weapons of Mass Destruction*, Hearings, Permanent Subcommittee on Investigations of the Committee on Governmental Affairs, United States Senate, Part II, March 13, 20, and 22, 1996, pp. 90–105.

12. Author's interview with American and Russian UN inspectors, Moscow, May 1997.

13. Ekeus. Also, "Testimony of David Kay, Senior Vice President, Hicks and Associates," *Global Proliferation of Weapons of Mass Destruction*, Hearings, Permanent Subcommittee on Investigations of the Committee on Governmental Affairs, United States Senate, Part II, March 13, 20, and 22, 1996, p. 325. Also, "Note by the Secretary General, to the U.N. Security Council," *Global Proliferation of Weapons of Mass Destruction*, Hearings, Permanent Subcommittee on Investigations, Committee on Governmental Affairs, United States Senate, Part I, October 31 and November 1, 1995, pp. 581–590. Also, Leonard A. Cole, *The Eleventh Plague*, W.H. Freeman, New York, 1996, p. 85.

14. Ibid.

15. *The 1997 GenCon Conference on Improving U.S. Capabilities for Defense from Bioterrorism*, January 16, 1997, The National Consortium for Genomic Resources Management and Services, Montross, Virginia, February 1997: Presentations by Scott Halstead, U.S. Naval Medical Research and Development Command (re: diarrhea), p. 21, and David Kay, SAIC, Inc. (re: testing on dogs), p. 34.

16. Author's interview with American UN inspector, Washington, D.C., May, 1997.

17. R. Jeffrey Smith, "Iraq's Drive for a Biological Arsenal," *The Washington Post*, November 21, 1997, p. A1.

18. See, for example, Milton Leitenberg, "Biological Weapons Arms Control," Center for International and Security Studies at the University of Maryland, College Park, Maryland, PRAC Paper No. 16, May 1996, p. 35.

19. Author's interview with American UN inspector, Washington, D.C., May 1997.

20. For a detailed description of Iraqi biochemical warfare facilities, see "Note by the Security General," S/1995/864, October 11, 1995, United Nations, included in *Global Proliferation of Weapons of Mass Destruction*, Hearings, Permanent Subcommittee on Investigations, Committee on Governmental Affairs, United States Senate, Part I, October 31 and November 1, 1995, pp. 581–590.

21. Smith, p. A1.

22. Iraqi Deputy Prime Minister Tariq Aziz interview, CNN News, January 14, 1998.

23. Leitenberg, pp. 23–39. Also, an unpublished report of November 1996 by Richard O. Spertzel of the UN Special Commission on Iraq reports: "By September 1996, the commission was monitoring eighty-six biological facilities throughout Iraq, including universities, breweries, food processing plants, and

production facilities for vaccines, antibiotics, biopesticides, and single cell protein (an animal feed supplement)."

24. Jonathan B. Tucker, "Chemical/Biological Terrorism: Coping with a New Threat," *Politics and the Life Sciences*, September 1996, p. 170.

25. Author's interviews with leading American geneticists, November 1996. See also, Elizabeth Pennisi, "First Genes Isolated from the Deadly 1918 Flu Virus," *Science*, March 21, 1997, p. 1739.

26. *The 1997 GenCon Conference*, Orientation Section, p. 2.

27. The extensive efforts of the Japanese cult Aum Shinrikyo to acquire biological as well as chemical agent capabilities are described in detail in many sections of the publication *Global Proliferation of Weapons of Mass Destruction*, Hearings, Permanent Subcommittee on Investigations, Committee on Governmental Affairs, United States Senate, Part I, October 31 and November 1, 1995. It appears, for example, that the cult sent a medical team to Zaire to obtain a sample of the Ebola virus during an outbreak in that country in 1992. Also, in late 1996 and early 1997, Promed Service posted information on the World Wide Web reporting the presence of Ebola in various monkey reservoirs in Africa.

28. Cole, *The Eleventh Plague*, p. 3.

29. Richard Preston, *The Hot Zone*, Anchor Books, Doubleday, New York, 1994.

30. Conflicting reports have been prepared about Japanese intentions and indeed actual use of biological agents in China. In any event, it seems clear that technical problems of delivering the agents were severe.

31. Tucker, p. 170.

32. Author's discussion with American scientists who had been engaged in military experiments, March, 1971.

33. Author's interview with American biological weapons experts, March, 1971.

34. Leitenberg, p. 21.

35. *Emerging Infections; Microbial Threats to Health in the United States*, Institute of Medicine, National Academy Press, Washington, D.C., 1992, pp. 126–127. Also, *The 1997 GenCon Conference*, Halstead, p. 20. Overseas laboratories were located in southeast Asia, northern Africa, sub-Saharan Africa, the Middle East, and on both sides of the South American continent.

36. Many reports are available about the American Type Culture Collection and its distribution policies. Congressional investigators have accused the organization of being indiscriminate in its exports of sensitive strains, even though the firm was technically complying with all export control requirements.

37. "Free Speech—and a Book That Goes Too Far," *The Washington Post*, November 16, 1997, p. C1.

38. *Medical Management of Biological Casualties*.

39. Author's interview with representative of Technical Steering Working Group on Counterterrorism, Department of State, March 1996.

40. Eliot Marshall, "Too Radical for NIH? Try DARPA," *Science*, February 7, 1997, p. 744. Also, "Pentagon-Funded Research Takes Aim at Agents of Biological Warfare," *The Journal of the American Medical Association*, August 6, 1997, p. 373.

41. William J. Cromie, "Turning Poisons into Vaccines," *Harvard University Gazette*, January 30, 1997, p. 1.

42. For future of vaccines see *The 1997 GenCon Conference*, Presentation of Anna Johnson-Winnegar, Department of Defense, p. 10.

43. *Protocol for the Prohibition of the Use in War of Asphyxiating, Poisonous or other Gases, and of Bacteriological Methods of Warfare,* available from the Department of State.
44. *Convention on the Prohibition of the Development, Production and Stockpiling of Bacteriological (Biological) and Toxin Weapons and Their Destruction,* available from the Department of State.
45. "Controlling Dangerous Pathogens: A Blueprint for U.S.–Russia Cooperation," National Academy of Sciences, October 25, 1997, Chapter 1.
46. *The Biological Weapons Convention,* Fact Sheet, The White House, January 27, 1998.
47. *The 1997 GenCon Conference,* Keynote Address by Elmo R. Zumwalt, President's Foreign Intelligence Advisory Board, p. 4.
48. For a discussion of the Australia Group, see *Assessing U.S. Efforts To Help Control Nuclear and Other Dangerous Materials and Technologies in the Former Soviet Union,* National Research Council, National Academy Press, Washington, 1997, Appendix A.
49. Leitenberg, p. 50.
50. Tucker, "Measures To Fight Chemical/Biological Terrorism," p. 245.
51. *The 1997 GenCon Conference,* Recommendations concerning the overlaps of terrorism and public health concerns of Fred Murphy, Dean of University of California (Davis) School of Veterinary Medicine, pp. 44–45.
52. See, for example, Karen Young Kreeger, "Smallpox Extermination Proposal Stirs Scientists," *The Scientist,* Vol. 10, No. 24, December 9, 1996, p. 11.
53. James R. Ferguson, "Biological Weapons and U.S. Law," *The Journal of the American Medical Association,* August 6, 1997, p. 357.
54. J.D. Simon, "Biological Terrorism: Preparing to Meet the Threat," *The Journal of the American Medical Association,* August 6, 1997, p. 97.

CHAPTER 5
THE GLOBALIZATION OF WEAPONS EXPERTISE

1. The Soviet system of forced labor camps dated back to 1919, and by the mid-1930s, the gulags housed several million inmates. According to stories in the Russian press during the fall of 1993, special battalions of prisoners were mobilized following World War II to construct the ten atomic cities. This activity apparently continued until the late 1950s at which time the gulags were scaled back, following Stalin's death. In 1995, Russian scientists from Chelyabinsk-70 acknowledged the use of prison labor there but were uncertain as to the duration of the use of such labor.
2. Grigory A. Yavlinsky, "Death of a Scientist," *The New York Times,* November 15, 1996, OpEd Page. Author's discussions in Moscow in November 1996 with close friends of Nechai. Also, David Hoffman, "Russian Turmoil Reaches Nuclear Sanctum, Suicide of Lab Director in 'Closed City' Underscores Angst," *The Washington Post,* December 22, 1996, p. A29.
3. The story of the Iraqi offer circulated in Moscow in 1992 but was never confirmed, at least publicly, as being true.

4. Author's visit to Chelyabinsk-70 in June 1994 and discussions in November 1996 with more recent American visitors to the city.

5. Hoffman, p. A29.

6. David E. Kaplan, "Terrorism's Next Wave," *U.S. News and World Report*, November 17, 1997, p. 26.

7. Glenn E. Schweitzer, *Moscow DMZ: The Story of the International Effort To Convert Russian Weapon Science to Peaceful Purposes*, M.E. Sharpe, Armonk, NY, 1996, p. 103. For more current information, see Richard Stone, "The Perils of Partnership," *Science*, January 24, 1997, pp. 468–471.

8. Schweitzer, p. 226.

9. Discussions with nuclear scientists involved in the design and operation of the Chernobyl atomic power station, Moscow and Obninsk, Russia, September 1992.

10. See for example, Masha Gessen, "The Future Ruins of the Nuclear Age," *Wired*, December 1997, p. 240.

11. Matt Taibbs, "Hero Missile Designer Shot Dead in Urals," *The Moscow Times*, March 22, 1996, p. 1. Also, author's discussions in Washington with officials of U.S. Department of Defense, December 1996.

12. Author's discussion with Department of Defense officials with oversight of activities in Russia, October 1996.

13. Author's discussions in 1993 and 1994 with Russian missile designers who displayed their wares at Abu Dabai during the early 1990s indicated that considerable Soviet technology was sold at bargain prices at the trade show.

14. Author's discussion with a leading Russian missile engineer indicated likely purchase of the missile by several countries, including Algeria, September 1996.

15. David Hoffman, "Russians Wrote Atomic History for Pentagon," *The Washington Post*, October 27, 1996, p. 1.

16. Author's discussions with Russian scientists from Sarov and Snezhinsk and with Department of Defense officials in Spring 1995 in Moscow.

17. Author's discussion with officials of the U.S. Department of Energy in Washington in December 1996.

18. Author's discussion with former Soviet weapons scientist who was denied access to Russian Ministry of Defense information center. Denial was related to the center's retrenchment after it became widely known it had been involved in dealings with American organizations operating under contract with the U.S. Department of Defense.

19. Previously documented in Schweitzer, pp. 69–70.

20. Ibid, p. 162.

21. Ibid, p. 53.

22. Ibid, pp. 151–152.

23. Ibid, p. 154.

24. Author's discussions in Moscow involving Russian scientists and Western assistance officials in Fall 1992.

25. Schweitzer.

26. See for example, *Proliferation Concerns*, National Research Council, National Academy Press, Washington, 1997.

27. These observations are based on author's discussions in Tbilisi with a variety of government officials and scientists in August 1994, May 1995, and October 1996.

28. John F. Sopko, *Global Proliferation of Weapons of Mass Destruction, Hearings,* Permanent Subcommittee of Investigations of the Committee on Governmental Affairs, U.S. Senate, Part II, March 13,10, and 22, 1996, p. 14.

29. Leaflet originally circulating in Middle East, obtained by German government in Spring 1995.

30. David Albright, "Engineer for Hire," *The Bulletin of the Atomic Scientists,* December 1993, pp. 29–36.

31. Ibid.

32. *World Military Expenditures and Arms Transfers,* 1995, U.S. Arms Control and Disarmament Agency, April 1996, pp. 12–13.

33. *International Military Education and Training,* Congressional Presentation for Foreign Operations, Fiscal Year 1996, United States Department of State, p. 194; *Foreign Military Sales, Foreign Military Construction Sales, and Military Assistance Facts,* Department of Defense Security Assistance Agency, September 30, 1995, pp. 93–113.

34. William J. Perry, Secretary of Defense, *Annual Report to the President and the Congress,* March 1996, p. 75.

35. *World Military Expenditures and Arms Transfers,* pp. 31–34.

36. Unpublished manuscript prepared by Marvin M. Miller, Center for International Studies, Massachusetts Institute of Technology, February 1998.

37. Ibid.

38. John W. Lewis and Xue Litai, *China Builds the Bomb,* Stanford University Press, Stanford, 1988.

39. For a discussion of Soviet efforts of Soviet technobandits, see Glenn E. Schweitzer, *Techno-Diplomacy, U.S.–Soviet Confrontations in Science and Technology,* Plenum, New York, 1989, Chapter 7.

40. During the 1980s and early 1990s, the National Academy of Sciences/National Research Council in Washington, D.C. produced a number of reports concerning the difficulties in controlling the international flows of sensitive technical data. Since that time, the introduction of Internet and the increase in the number of countries with advanced technological capabilities have made these flows are even more difficult to address.

41. Author's discussions with nuclear weapons designers at Sandia National Laboratory, Albuquerque, New Mexico, March 1996.

42. For a general discussion of the dual-use problem, see ibid, Chapter 6.

43. Department of State Briefing, July 1996.

CHAPTER 6
DRUG RUNNERS AND TERRORISTS: NEW PARTNERS IN CRIME

1. Kevin Fedarko, "The Capture of America's Most Wanted," *Time,* January 29, 1996, pp. 50–51.

2. Abrego was subsequently sentenced to eleven life terms, fined $128 million, and stripped of $350 million in assets as reported in *International Narcotics Control Strategy Report, 1996,* Department of State, March 1997, p. 1.

3. Martha Brant, "Liposuctioned to Death," *Newsweek*, July 21, 1997, p. 43. Molly Moore, "Dead Drug Lord's Doctors Founded Embedded in Concrete," *The Washington Post*, November 7, 1997, p. A29. Further confusing the situation, conflicting reports have been received as Mexican efforts to mask what has happened to the surviving surgeons. See, for example, Douglas Farrah and Molly Moore, "Doctor in Drugland's Operation Sheltered in U.S.," *The Washington Post*, April 11, 1998, p. A05.
4. Molly Moore, "U.S. Names Mexican Trafficker to FBI's Wanted List," *The Washington Post*, September 25, 1997, p. A29.
5. Lupsha, p. 43.
6. Robert J. Nieves, *Colombian Cocaine Cartels, Lessons from the Front*, National Strategy Information Center, Washington, April 1997, p. 25.
7. Ibid. Also for estimated costs, see Douglas Farah, "Mexican Control of U.S. Cocaine Market Grows," *The Washington Post*, August 5, 1997, p. A11.
8. For a discussion of links in a variety of countries between drug dealers and terrorists, see the following: Richard Clutterbuck, *Terrorism in an Unstable World*, Routledge, London and New York, 1994, chapters 9 and 12; Peter Lupsha, "Transnational Organized Crime Versus the Nation State," *Transnational Organized Crime*, Vol. 2, Spring 1996, Number 1, p. 28; Senator John Kerry, *The New War*, Simon and Schuster, 1997, p. 29; and Alex P. Schmid, "The Link Between Transnational Organized Crime and Terrorist Crimes," *Transnational Organized Crime*, Vol. 2, No. 4. Winter 1996, p. 63.
9. *International Narcotics Control Strategy Report*, p. 8. See also, comments by General Barry R. McCaffrey, Director of the Office of National Drug Control Policy, made before West Point Society, Fort Myer Officers Club, January 14, 1998.
10. See sources cited in footnote 3.
11. Ibid.
12. Douglas Farah, "Russian Mob, Drug Cartels Joining Forces," *The Washington Post*, September 29, 1997. See also, Kerry.
13. Schmid, p. 40. See also, Kerry, p. 31.
14. *International Narcotics Control Strategy Report, 1996*, p. 2.
15. *Larry King Live*, CNN, "Interview with Ernesto Samper," December 9, 1997.
16. Glenn C. Schoen, Lecture on Narco-Terrorism at Georgetown University, Washington, D.C., March 20, 1997, presenting extensive evidence concerning the synergism between drug cartels and guerrillas/terrorists.
17. William Branigin, "Drug Gangs Terrorize the Texas Border," *The Washington Post*, September 25 1996, p. A1. Also, see Douglas Farrah and Molly Moore, "2000 Miles of Disarray in Drug War," *The Washington Post*, March 9, 1998, p. A5.
18. See, for example, John Sweeney, "In Latin America, Quelling Chaos," *Foreign Service Journal*, October 1996, pp. 34–39. For update, see Pamela Falk, "Drugs Across the Border; A War We're Losing," *The Washington Post*, September 24, 1997, p. A21.
19. Linda Robinson, "An Inferno Next Door," *Time*, February 24, 1997, p. 36.
20. Molly Moore, "Two Mexican Prison Officials Resign," *The Washington Post*, April 10, 1997, p. A27.
21. *International Narcotics Control Strategy Report, 1996*.
22. Ibid, p. 20. See also, Kerry, p. 82.
23. See, for example, Jess T. Ford, "Drug Control, U.S. Heroin Control Efforts in Southeast Asia," United States General Accounting Office, GAO/T-NSIAD-96-

240, September 19, 1996; "Heroin, Facts and Figures," Office of National Drug Control Policy, PK 26, January 1996; and Alfred McCoy, "In Asia, Battling Warlords," *Foreign Service Journal*, October 1996, pp. 28–33.

24. Author's interview with chairman of Georgian Parliamentary Commission on International Affairs, November 1996.

25. Sandra Dallas (editor), "A Tragic Growth Industry: Drugs," *Business Week*, June 9, 1997, p 42.

26. Lynne Duke, "Drug Trade Moves in on South Africa," *The Washington Post*, September 1, 1994, p. A33.

27. Howard Schneider, "Vancouver Boom Has a Down Side," *The Washington Post*, April 24, 1997, p. A27.

28. CNN Television Series, *Impact*, October 13, 1997. While Emory's store HEMPBC has been closed several times for brief periods, he has always succeeded in reopening within a few days while the Mounties concentrate on more dangerous criminals.

29. Harry W. Shlaudeman and W. Kenneth Thompson, "In Washington, Maintaining Pressure," *Foreign Service Journal*, October 1996, p. 21; a "Clandestine Laboratories and Precursor Chemicals," National Institute of Justice, TS11666/9509, 1995; and Robert Suro, "Other Drugs Supplanting Cocaine Use," *The Washington Post*, June 25, 1997, p. A1.

30. Shlaudeman, p. 21.

31. *The National Drug Control Strategy*, 1997, The White House, February 1997, p. 63.

32. Raphael F. Perl, "Drug Control: International Policy and Options," CRS Issue Brief, Order Code IB88093, p. 11.

33. Stephen Rosenfeld, "Drug War: The Enemy Within," *The Washington Post*, March 7, 1997, p. A23.

34. Perl, p. 11.

35. *International Narcotics Control Strategy Report*, p. 6. The report begins on a note of high optimism. At about the same time, the press reported the lack of success of efforts of the Clinton Administration to put a positive spin on the president's visit to Mexico in Spring 1997. See, for example, Peter Baker and John Ward Anderson, "Clinton: U.S. Shares Blame with Mexico for Drug Trade," *The Washington Post*, April 7, 1997, pp. A1.

36. Shlaudeman, p. 22.

37. Senator Charles Grassley, chairman of the Senate Caucus on Narcotics Control, made similar remarks during a speech at the Heritage Foundation Conference on *International Corruption*, Washington, March 20, 1997. For a more detailed discussion of the overall networks of the cartels, see Sidney Jay Zabludoff, "Colombian Narcotics Organizations as Business Enterprises," *Transnational Organized Crime*, Vol. 3, No. 2, Summer 1997, pp. 30–50.

38. Kerry, p. 78 and 80.

39. Kerry, p. 87.

40. *International Narcotics Control Strategy Report*.

41. Ibid, p. 2.

42. Department of State Briefing, "Transnational Crime Seminar," Overseas Security Advisory Council, November 6, 1997.

43. Martha Brant, "A Defector in the Drug War," *Newsweek*, March 3, 1997, p. 54.

44. "General Praises Colombia Drug War," *The Washington Post*, March 7, 1997, p. A12.

45. Charles Bowden, "The Killer Across the River," *GQ*, April 1997, p. 212.

46. "U.S. Envoy Says Drug Lords Offered Evidence on Samper," *The Washington Post*, March 7, 1997, p. A12.

47. Geneva Overholser, "The CIA, Drugs, and the Press," *The Washington Post*, November 10, 1996, p. C6. Also, Steven Holmes, "Accusations of CIA Ties to Drug Ring Are Reviewed," *The New York Times*, April 15, 1997, p. A30.

48. Michael R. Bromwich, Inspector General, "Statement," Caucus on International Narcotics Control, U.S. Senate, May 14, 1997, Available from Department of Justice, May 14, 1997.

49. Ibid. Also, Dan McGraw, "18 Law Enforcement Officials Convicted in Border-Related Drug Corruption," *US News and World Report*, February 24, 1997, p. 40.

50. Molly Moore, "Mexican Airborne Anti-Drug Unit Arrested in New Cocaine Scandal," *The Washington Post*, September 11, 1997, p. A27.

51. Glenn Frankel, "U.S. War on Drugs Yields Few Victories," *The Washington Post*, June 8, 1997, p. A1.

52. Information provided by the Office of National Drug Control Policy, October 1997.

53. See *International Narcotics Control and United States Foreign Policy: A Compilation of Laws, Treaties, Executive Documents and Related Material*, Senate Caucus on International Narcotics Control, Committee on International Relations U.S. House of Representatives, September 1997.

54. *Drug Control*, "Update on U.S. Interdiction Efforts in the Caribbean and Eastern Pacific," General Accounting Office, GAO/NSIAD-98-30, October 1997, p. 15.

55. *The National Control Drug Strategy, 1996: Program, Resources, and Evaluation*, pp. 241–245.

56. Ibid.

57. Douglas Farah and Serge F. Kovaleski, "Cartels Make Puerto Rico a Major Gateway to the U.S." *The Washington Post*, February 16, 1998, p. A1.

58. Ibid, pp. 86–91. Also, Bradley Graham, "McCaffrey Wants Pentagon To Spend More Against Drugs," *The Washington Post*, November 7, 1997, p. A3; and Dana Priest, "U.S. May Boost Military Aid to Colombia's Anti-Drug Effort," *The Washington Post*, March 28, 1998, p. A19.

59. "International Drug Trade and the U.S. Certification Process: A Critical Review," Caucus on International Narcotics Control, the United States Senate, September 1996, pp. 1–2.

60. Bowden, p. 210. Also, Douglas Farah, "Mercenaries at Work for Mexico's Drug Families," *The Washington Post*, October 30, 1997, p. A25; Douglas Farah, "U.S. Aid in Limbo as Colombian Army Fails To Provide Evidence on Rights Abuse," *The Washington Post*, January 10, 1998, p. A20; *International Narcotics Control Strategy Report*, U.S. Department of State, March 1998, pp. xxv and xxxii.

61. State Department Briefing, "Transnational Crime Seminar."

62. Shlaudeman, p. 22; and Sweeney, p. 35.

63. Pierre Thomas, "Drug Dealers Employ U.S. Mail as Carrier," *The Washington Post*, February 15, 1997, p. A1.

CHAPTER 7
LAUNDERED ASSETS:
CLEANING DIRTY CURRENCY TO FUEL
THE TERRORIST ENGINE

1. Linda P. Raine and Frank J. Cilluffo (editors), *Global Organized Crime and the New Evil Empire*, Center for Strategic International Studies, Washington, D.C., comments by Jack Blum, pp. 22–23.
2. "Financial Crimes and Money Laundering," International Narcotics Control Strategy Report, 1996, Department of State, March 1997.
3. Linda Robinson and Doug Pasternak, "Rum Sort of Banking," *U.S. News and World Report*, October 20, 1996, pp. 38–41.
4. For general background, see Jack A. Blum, *Enterprise Crime: Financial Fraud in International Interspace*, National Strategy Information Center, Washington, June 1997.
5. "Global Proliferation of Weapons of Mass Destruction," *Hearings*, Permanent Subcommittee on Investigations, Committee on Governmental Affairs, U.S. Senate, Part I, October 31 and November 1, 1995, pp. 57–58.
6. "Financial Crimes and Money Laundering." Also, author's interview with Department of State counterterrorism official, March 1997.
7. Robinson and Pasternak.
8. Richard Clutterbuck, *Terrorism in an Unstable World*, Routledge, London and New York, 1994, p. 129.
9. *Global Organized Crime and the New Evil Empire*, remarks by Senator Patrick Leahy, pp. 18–19.
10. Ibid.
11. Ibid.
12. *Global Organized Crime*, remarks by Robert Roser, p. 21.
13. Speech by General Barry R. McCaffrey, Special Assistant to the President for Narcotics Control, World Affairs Council, Washington, D.C., March 26, 1997.
14. Ibid.
15. David A. Andelman, "The Drug Cartel's Weak Link," *Foreign Affairs*, July/August 1994, p. 94.
16. Jeffrey Robinson, *The Laundrymen*, Arcade, New York, 1996, p. 37.
17. Robert D. McFadden, "Limits on Cash Transactions Cut Drug-Money Laundering," *New York Times*, March 11, 1997, p. 1.
18. Ibid.
19. Andelman, p. 98.
20. Ibid, p. 101.
21. *Global Organized Crime and the New Evil Empire*, remarks by William McLucas, pp. 17–18.
22. Andelman, p. 101.
23. *Global Organized Crime and the New Evil Empire*, remarks by Don Parker, p. 42.
24. *Information Technologies and Money Laundering*, Office of Technology Assessment, U.S. Congress, Government Printing Office OTA-ITC-630, September 1995, pp. 1–2.
25. Ibid, p. 23.

26. Ibid, Chapter 1.
27. John Kerry, *The New War: The Web of Crime that Threatens America's Security,* Simon and Schuster, New York, 1997, p. 158.
28. Robinson, pp. 183–184.
29. "Financial Crimes and Money Laundering," p. 34, WWW, January, 1998.
30. Ibid.
31. *Global Organized Crime and the New Evil Empire,* Blum, p. 33.
32. Ibid., p. 28. For a list of Commonwealth countries, including British Dependencies, see "Independent Countries and Dependencies,"http://www.britain. org.za/britain/commonwealth.html February 1998.
33. *Global Organized Crime and the Evil Empire,* pp. 24–25.
34. Douglas Farah, "Antigua Internet Bank Vanishes into Cyberspace," *The Washington Post,* August 31, 1997, p. A30.
35. Ibid.
36. Ibid.
37. "Cleaning Up Dirty Money," *The Economist,* August 26, 1997, p. 13. Another case study is presented in detail in Tom Blickman, "The Rothschilds of the Mafia on Aruba," *Transnational Organized Crime,* Vol. 3, No. 2, Summer 1997, pp. 50–90.
38. Robinson, p. 4.
39. Ibid, pp. 4–9.
40. Ibid.
41. Ibid, pp. 20–22.
42. Ibid, pp. 21–22.
43. Sharon Walsh, "Clifford, Altman Settle BCCI Case," *The Washington Post,* February 4, 1998, p. A1.
44. *Money Laundering: U.S. Efforts To Fight Are Threatened by Currency Smuggling,* General Accounting Office, GAO/GGD-94-73, March 9, 1994.
45. Form Letter Distributed by Fax from Dr. Abe Odum, Lagos, Nigeria, October 21, 1996. For a detailed discussion of this type of Nigerian fraud, see the report on hearings of the House International Relations Committee, Subcommittee on Africa, in "Documentation" *Transnational Organized Crime,* Vol. 3, No. 2, Summer 1997, pp. 131–149.
46. Ibid.
47. See, for example, "Central Bank of Nigeria," advertisement, *The Washington Post,* September 3, 1997, p. A7.
48. Robinson, p. 273.
49. *International Narcotics Control Strategy Report,* pp. 32–52 on World Wide Web.
50. Ibid.
51. "Cleaning Up Latin America," *The Economist,* April 6, 1996, p. 41.
52. *International Narcotics Control Strategy Report,* p. 24 on World Wide Web.
53. Andelman, p. 101.
54. Robinson, p. 307.
55. "Following the Money in Cyberspace," background report of the Financial Crimes Enforcement Network, Washington, September 1997.
56. Raine and Cilluffo, remarks by Stanley Morris, p. 62.
57. Ibid.
58. See, for example, "Money Laundering" *The Economist,* July 26, 1997, pp. 19–21.
59. Kerry, p. 157.
60. Walsh, p. A1.

CHAPTER 8
BUSINESS AS USUAL:
HIJACKINGS, BOMBINGS, AND ASSASSINS

1. Alice Reid and Peter Pae, "Security Breach Brings Dulles Airport to a Halt," *The Washington Post*, June 12, 1997, p. D1; Jennifer Ordonez, "Bomb Scare Empties D.C. Building Again," *The Washington Post*, June 19, 1997, p. A6.
2. Serge Kovalevski and Pierre Thomas, "Intelligence Network Fails To Link Crash to Terrorism," *The Washington Post*, September 14, 1996, p. A3; "Clues from the Sky," *U.S. News and World Report*, July 29, 1996, pp. 22–25.
3. Dale Russakoff, "Deliberations Begin in Jet Bomb Plot Case," *The Washington Post*, August 30, 1996, p. A3. Also, "Yousef Sentenced to 240 Years in Solitary," *USA Today*, January 9, 1998, p. 1A.
4. See, for example, Keith O. Fultz, "Aviation Security; Urgent Issues Need to Be Addressed," General Accounting Office, AGAO/T-RCED/NSIAD-96-251, September 1996.
5. Ibid.
6. "Criminal Acts Against Civil Aviation, 1995," Office of Civil Aviation Security, Federal Aviation Administration, 1996.
7. Ibid, p. 15.
8. Peter St. John, *Air Piracy, Airport Security, and International Terrorism*, Quorum Books, Westport, Connecticut, 1991, Appendix 2.
9. Richard Clutterbuck, *Terrorism in an Unstable World*, Routledge, New York and London 1994, pp. 169–170.
10. *Airline Passenger Security Screening*, National Research Council, National Academy Press, Washington, 1996, p. 6.
11. St. John, 70–71.
12. John Mintz, "Gore Panel Proposes Aviation Safety Steps," *The Washington Post*, February 13, 1997, p. C1.
13. Ibid.
14. Greg Nojeim, "Profiles in Pointlessness," *The Washington Post*, January 27, 1997, OpEd page. Keith Alexander, "Fliers Protest Profiling," *USA Today*, September 30, 1997, p. 10B.
15. St. John, Appendix 6.
16. Ibid, p. 71.
17. Keith O. Fultz, "Aviation Security; Technology's Role in Addressing Vulnerabilities," General Accounting Office, GAO/T-RCED/NSIAD-96-262, September 1996.
18. *Airline Passenger Security Screening*, p. 15.
19. Ibid; also, Fultz, "Aviation Security: Technology's Role."
20. St. John, p. 71.
21. Mintz.
22. David Kaplan, "Bomb-sniffing Tests Provoke a Dogfight," *U.S. News and World Report*, November 24, 1997, p. 42.
23. "Unlocking the Secrets of Supersniffing Dogs," *Chemical and Engineering News*, September 29, 1997.
24. Clutterbuck, pp. 55–63.
25. Roger Medd and Frank Goldstein, "International Terrorism on the Eve of a New Millennium," *Studies in Conflict and Terrorism*, July–September, 1997, p. 298.

26. Pierre Thomas, "Vast Plastic Explosives Supply Cited," *The Washington Post*, January 21, 1997, p. A6.

27. Robert F. Service, "NRC Panel Enters the Fight over Tagging Explosives," *Science*, January 24, 1997, pp. 474–475.

28. Clutterbuck, p. 24.

29. "Irish Bombs Temper Investors' Talk," The Associated Press, October 8, 1996.

30. "Bombings in France, 1995," *The Washington Post*, December 4, 1996, p. 6.

31. Nesha Starcevic, "Death Strikes as Song Fills Church," *USA Today*, December 26, 1996, p. 6A.

32. William Drozdiak, "Eurobomber Case Stumps Austrian Police," *The Washington Post*, November 17, 1996, p. A24.

33. Andrew MacDonald, *Turner Diaries: A Novel*, Barricade Books, New York, Second Edition, 1996.

34. *The Protection of Federal Office Buildings Against Terrorism*, National Research Council, National Academy Press, Washington, 1988.

35. "1995 Bombing Incidents," (with supplements), FBI Explosives Unit-Bomb Data Center, 1996.

36. Ordonez.

37. "Man Blows His Hand Off in Booby-Trapped Apartment," *The New York Times*, May 2, 1997, p. B4.

38. "Terrorism Advisory Called Off by FBI After Truck Is Found," *The Washington Post*, February 26, 1997, p. A11.

39. "Arab Paper Receives Letter Bombs in London, U.N.; 2 Hurt," *The Washington Post*, January 14, 1997, p. A9.

40. Author's discussions with expert responsible for National Research Council report in-progress on the topic of explosives, January 1998.

41. Ibid.

42. Ibid.

43. Bruce W. Nelan, "How They Did It," *Time*, May 5, 1997, pp. 57–61.

44. James Nixey, "Journalists Released After Chechen Ordeal," *Moscow Tribune*, June 7, 1997, p. 1.

45. Lee Hockstader, "Journalists Become Chechnya's Latest Victims," *The Washington Post*, May 27, 1997, p. A10. "Chechnya, Habeas Corpus," *The Economist*, August 23, 1997, p. 42.

46. Sue Anne Pressley, "Separatist in Texas Takes Two Hostages Blaming Police for Kidnapping Associates," *The Washington Post*, April 28, 1997, p. A6; Sam Howe Verhover, "Leader of Armed Group Breaks Off Talks on Settling Standoff in West Texas," *The Washington Post*, May 1, 1997, p. A16.

47. Gordon Witkin, Peter Cary, and Linda Robinson, "Peru Takes No Prisoners," *U.S. News and World Report*, May 5, 1997, pp. 32–34.

48. Clutterbuck, p. 177.

49. Karen Kresbach, "Surviving Terror in Peru," *Foreign Service Journal*, February 1997.

50. Adrian Havill, "Line of Fire," *The Washingtonian*, August 1997, p. 73.

51. See, for example, Robert O'Harrow, Jr, "Suspect in CIA Slayings Is Returned to U.S.," *The Washington Post*, June 18, 1997, p. 1; Wendy Melillo, "Suspect Puts Fairfax in Judicial Spotlight," *The Washington Post*, June 25, 1997, p. A8; "2 Governments Cloak Details of the Capture," *The Washington Post*, June 19, 1997, p. A10; "The Very Long Arm of the Law," *The Washington Post*, June 19, p. A20. Brooke Masters and Wendy Melillo, "FBI Had Foreign Help in CIA

Suspect's Arrest," *The Washington Post*, October 21, 1997, p. B5; Wendy Melillo and Peter Finn, "Mir Aimal Kasi Likely To Receive Death Sentence Today in Killings Outside CIA," *The Washington Post*, January 23, 1998, p. A18.

52. Howard Schneider, "A Journey into Violence," *The Washington Post*, August 5, 1997, p. A10.

53. "The Cold War: Fifty Years of Silent Conflict Exhibit," *What's New at CIA*, CIA, October 1997, pp. 33–34.

54. "PCCIP: Background," President's Commission on Critical Infrastructure Protection, January 1997, Office of the Press Secretary, The White House.

55. "Critical Foundations: Protecting America's Infrastructure," The Report of the President's Commission on Critical Infrastructure Protection, October 1997.

56. Michael S. Serrill, "Subway Scare: Terror Takes Aim at New York," *Time*, August 11, 1997, p. 36.

CHAPTER 9
THE GROWING ARSENAL
TO COMBAT TERRORISM

1. "Designation of Terrorist Organizations," Briefing by Secretary of State, Washington, D.C., October 8, 1997.

2. Ibid.

3. For an example of practical difficulties in implementing new legislative provisions see William Claiborne, "New Antiterrorism Law Suffers Legal Setbacks," *The Washington Post*, July 11, 1997, p. A17.

4. Greg McCullough, "Section-by-Section Summary of Antiterrorism Bill," Legislative News Service, Washington, D.C., April 15, 1996; this is a report on legislation signed on April 24, 1996.

5. Raphael F. Perl, "Terrorism, the Future, and U.S. Foreign Policy," CRS Issues Brief, Congressional Research Service, July 11, 1997.

6. For a cynical appraisal of the cost effectiveness of these expanding activities see David E. Kaplan, "Everyone Gets into the Terrorism Game," *U.S. News and World Report*, November 17, 1997, p. 32.

7. *Patterns of Global Terrorism, 1996*, United States Department of State, April 1997, pp. iii–vi, and "President's National Security Strategy," April 1997, available from White House Press Office, April 1997.

8. See, for example, discussions of activities in Middle East in Stansfield Turner, *Terrorism and Democracy*, Houghton Mifflin, Boston, 1991. Also, see Perl, p. CRS-3.

9. John Deutch, "Terrorism," *Foreign Policy*, Fall, 1997.

10. *International Terrorism: A Compilation of Major Laws, Treaties, Agreements, and Executive Documents*, Prepared for the Committee on Foreign Affairs, U.S. House of Representatives, Congressional Research Service, December 1994.

11. "Counterterrorism Fact Sheet," Office of the Press Secretary, The White House, April 30, 1996.

12. "Ministerial Conference on Terrorism: Agreement on 25 Measures," Text of Agreement released at the Ministerial Conference on Terrorism, Paris, France, July 30, 1996, available from Department of State, beginning at that date.

13. "Final Communique, Summit of the Eight," Office of the Press Secretary, The White House, June 22, 1997.

14. *Proliferation: Threat and Response,* "The Transnational Threat," Department of Defense, November 1997.

15. An overview discussion of the Nonproliferation Treaty and the role of the IAEA is contained in *Proliferation Concerns: Assessing U.S. Efforts To Help Contain Nuclear and Other Dangerous Materials and Technologies in the Former Soviet Union,* National Research Council, National Academy Press, Washington, 1997. See, in particular, Appendix 1. Many more detailed discussions are available from the International Atomic Energy Agency in Vienna, Austria, and from the Office of Arms Control and Nonproliferation of the Department of Energy.

16. Ibid. This source also discusses other export control regimes. Additional information is available from the U.S. Department of Commerce, Bureau of Export Control.

17. Ibid. Also, *Proliferation,* for discussion of capabilities of individual countries.

18. "Response to Threats of Terrorist Use of Weapons of Mass Destruction," Report of the President to Congress, January 31, 1997.

19. Ibid. See also, "Terrorism, Weapons of Mass Destruction, and U.S. Security," *The Monitor,* Center for International Trade and Security, University of Georgia, Summer 1997, p. 25; and for a listing of laws calling for sanctions, see Dianne E. Rennack and Robert D. Shuey, "Economic Sanctions to Achieve U.S. Foreign Policy Goals: Discussion and Guide to Current Law," *CRS Report to Congress,* #97-949, October 20, 1997.

20. "Fact Sheet, Iran-Libya Sanctions Act of 1996," Office of the Press Secretary, The White House, August 5, 1996.

21. Perl, p. CRS-5.

22. Robert A. Pape, "Why Economic Sanctions Do Not Work," *International Security,* Fall 1997, p. 90.

23. *Patterns of Global Terrorism,* 1996, Department of State, April 1997, p. 22.

24. Perl, p. CRS-11.

25. Author's interview with officials of Department of State, July 1997.

26. Lee H. Hamilton, "Speech," BXA's 10th Annual Update Conference, July 8, 1997, Released by Department of Commerce, July 10, 1997.

27. Paul Craig Roberts, "A Growing Menace to Free Trade: U.S. Sanctions," *Business Week,* November 24, 1997, p. 28.

28. Hamilton; Thomas Lippman, "U.S. Rethinking Economic Sanctions," *The Washington Post,* January 26, 1998, p. A6; Barton Gellman, "Shift in Iran Fuels Debate over Sanctions," *The Washington Post,* December 31, 1997, p. A1; for a particularly skeptical appraisal of the impact of economic sanctions, see Pape.

29. "Hearing on Iran and the Proliferation of Weapons," Senate Committee on Foreign Relations, Subcommittee on Near Eastern and South Asian Affairs, April 17, 1997, transcript from Federal Document Clearing House, Inc., Washington, D.C., April 1997.

30. Ibid, testimony of David Welch, Department of State.

31. Ibid. Also see Paul Geitner, "German Court: Iran Ordered Killing," Associated Press Report, April 10, 1997; and William Drozdiak, "Iranian Group Threatens Bombings Unless Germans Apologize for Court Ruling," *The Washington Post,* April 19, 1997, p. A17.

32. "Hearings on Iran," testimony of Robert Einhorn, Department of State.

33. The International Atomic Energy Agency, headquartered in Vienna, Austria, and the associated informal Nuclear Suppliers Group devote considerable international resources and time to limiting the diffusion of sensitive technologies to irresponsible states.

34. Thomas W. Lippman, "U.S. Keeps After Russia to Halt Flow of Missile Technology to Iran," *The Washington Post*, January 18, 1998, p. A9.

35. Melinda Liu and Christopher Dickey, "A Soft Signal from Iran," *Newsweek*, January 19, 1998, p. 30.

36. *Proliferation*.

37. Author's discussions with State Department officials, July 1997, and "Prepared Statement of Louis J. Freeh," Hearing on Counterterrorism, Senate Appropriations Committee, May 13, 1997, available from Federal News Service, Washington, May 1997.

38. "Terror Threat Home-Grown, Officials Say," Associated Press Report in *USA Today*, May 14, 1997; David E. Kaplan and Mike Tharp, "Terrorism Threats at Home," *U.S. News and World Report*, December 29, 1997/January 5, 1998, p. 22.

39. Freeh.

40. Ibid.

41. Ibid.

42. "Prepared Statement of Janet Reno," Hearing on Counterterrorism, Senate Appropriations Committee, May 13, 1997, available from Federal News Service, Washington, May 1997.

43. Testimony of Louis Freeh before the Subcommittee on Technology, Terrorism, and Governmental Affairs, Judiciary Committee of the U.S. Senate, September 3, 1997. Considerable background information on this issue is contained in *Cryptography's Role in Securing the Information Society*, National Research Council, National Academy Press, Washington, 1996.

44. Freeh, September 3, 1997.

45. Ibid.

46. Dorothy E. Denning and William E. Baugh, *Encryption and Evolving Technologies: Tools of Organized Crime and Terrorism*, National Strategy Information Center, Washington, July 1997.

47. Testimony by panel of business representatives before the Subcommittee on Technology, Terrorism, and Governmental Affairs, Judiciary Committee of the U.S. Senate, September 3, 1997.

48. Kenneth Lieberman, Vice President of Visa Corporation, Testimony before the Subcommittee on Technology, Terrorism, and Governmental Affairs, Judiciary Committee of the U.S. Senate, September 3, 1997.

49. Rajiv Chandrasekaran, "Freeh Seeks Encryption Decoding Key," *The Washington Post*, September 4, 1997, p. E1.

50. Ibid.

51. Todd Lappin, "Too Close for Comfort," *Wired*, December 1997, p. 122.

52. *Emergency Response to Terrorism, Self Study*, U.S. Department of Justice, Federal Emergency Management Agency, FEMA/USFA/NFA-ERT:SS, p. 16.

53. Ibid, p. 22.

54. Ibid, p. 23.

55. Edward Plaugher, "Washington Metropolitan Medical Strike Team," Presentation to National Research Council, Washington, D.C., July 1997.

56. Ibid.

57. Ibid.

58. Ibid.
59. Ibid.

CHAPTER 10
SEARCHING FOR LONG-TERM
SOLUTIONS TO REPLACE QUICK FIXES

1. Adapted from the following: unpublished paper by Phil Williams and Paul Woessner, "Transnational Threats and European Security," June 1997, University of Pittsburgh; Terry N. Mayer, "Battlefield of the Future," Chapter 9, on Internet at sdsar.af.mil/battle/chp8.html, August 1997; the "GenCon Conference on Improving U.S. Capabilities for Defense from Bioterrorism," GenCon, Washington, D.C., January 16, 1997; and comments by Gideon Frieder at George Washington University seminar "Terrorism and Civil Defense," August 12, 1997.
2. Raphael Perl, "Terrorism, the Future, and Foreign Policy," CRS Issues Brief, IB95112, Congressional Research Service, February 19, 1997.
3. Meeting among Senators and staff in Washington. D.C., as reported by a participant in the meeting, June 1997.
4. *Patterns of Global Terrorism 1996*, United States Department of State, April 1997. For a relevant discussion of definitional issues, see Maratha Crenshaw and John Pimlott (Editors), *Encyclopedia of World Terrorism*, M.E. Sharpe, Armonk, NY, 1997, Vol. 1, pp. 12–23.
5. As to allegations concerning the Algerian Army, televised report on CBS, *Sixty Minutes*, Christiana Amanpour report, January 18, 1998.
6. For related views, see Walter Lacquer, "Post Modern Terrorism," *Foreign Affairs*, September/October 1997, p. 24–36.
7. Gordon Gray, Department of State, at a seminar "International Cooperation in Combatting Terrorism," sponsored by George Washington University, Washington, D.C., February 25, 1998.
8. Phil Williams, "Getting Rich and Getting Even," Background Paper for Global Organized Crime Project, Center for Strategic and International Studies, Washington, D.C., 1996. Also, Alex P. Schmid, "The Links between Transnational Organized Crime and Terrorist Crimes," *Transnational Organized Crime*, Volume 2, Number 4, Winter 1996, pp. 40–82.
9. Phil Williams put forth the idea of hybrids at a meeting at the National Research Council, Washington, D.C., July 24, 1997.
10. See for example, Robert Sure, " '2 Terrorist Groups Set Up U.S. Cells,' Senate Panel Is Told," *The Washington Post*, May 14, 1997, p. A4.
11. *Patterns of Global Terrorism*, p. vi.
12. Senator John Kerry, *The New War*, Simon and Schuster, New York, 1997, p. 168.
13. For a discussion of coordination of governmental programs see, *Combating Terrorism*, GAO/NSIAD-98-39, December 1997.
14. During discussion of fast tracking processing for pre-cleared passengers based on submission of "intrusive" personal data, an American professor from the University of Wisconsin who is a specialist in civil liberties contended that such "coercion" to provide personal data would be discriminatory for those

people who did not want to sacrifice personal liberties; and therefore the practice should not be instituted, December 29, 1997. However, the logic of penalizing one group because of objections of another to the behavior of the first group seems to distort the concept of civil liberties.

15. See, for example, David Montgomery, "*Hit Man* Book Not Protected Under Free Speech, Lawyers Say," *The Washington Post,* May 6, 1997, p. D1.

16. Mark Hansen, "No Place To Hide," *ABA Journal,* August 1997.

17. Aziz Abu Hamad, "Death by Deportation," *The Washington Post,* May 19, 1997. This Op Ed contribution raises the issue of the amount of certainty of guilt needed for deportation to countries where the deportee may not receive a fair trial.

18. For discussions of many of the legal and related issues, see Alfred Rubin, "Current Legal Approaches to International Terrorism: Alternative Legal Approaches," Harry H. Almond, "The Legal Regulation of International Terrorism: Alternatives and Recommendations," and Oscar Schachter, "The Lawful Use of Force by a State Against Terrorists in Another Country," all in Henry H. Hahn (editor), *Terrorism and Political Violence: Limits and Possibilities of Legal Control,* Oceana Publications, New York, 1990. Also, for another dimension see Jack Donnelly, "Rethinking Human Rights," *Current History,* November 1996, pp. 387–391.

19. For an interesting perspective on terrorism and civil liberties, see Benjamin Netanyahu, *Fighting Terrorism,* Noonday New York, 1997.

20. See, for example, "Under the Microscope," *Time,* April 28, 1997, pp. 28–34; "U.S. Border Patrol Agent Accused of Smuggling Drugs," *The Washington Post,* August 7, 1997, p. A20; and Jeff Stein, "Spies and Sickos," on the Internet at Newsreel home page in May 1997.

21. For a popular perspective, see Brock Brower, "Can We Trust the FBI," *Readers Digest,* August 1997, pp. 77–81.

22. Since 1984, the Department of State has had in place a program offering rewards of up to $4 million for information leading to the arrest or conviction of terrorists who target U.S. citizens or property. Also, a special program that provides for rewards for information concerning nuclear materials has been in place for many more years. See, "Counter Terrorism Rewards Program," Department of State HEROES Home page on Internet, August 1997.

23. *Congressional Record,* June 27, 1996, pp. S7076–S7077. This report by Senator John Glenn discusses The *Atomic Weapons and Special Nuclear Materials Rewards Act of 1955,* the *Nuclear Proliferation Prevention Act of 1994,* and the Nunn-Lugar-Domenici legislation of 1996.

24. Author's discussions in April 1997 with Department of State officials who had met with the Fertilizer Institute leaders immediately following the Oklahoma City incident.

25. George Tenet, "The Intelligence Community's Role in Combating Terrorism," Testimony before the Senate Appropriations Committee, May 13, 1997, available from Federal News Service, Washington.

26. Ibid.

27. "The Acquisition of Technology Relating to Weapons of Mass Destruction and Advanced Conventional Munitions," Director of Central Intelligence, June 1997. The Monterey Institute of International Studies, for example, maintains an extensive data base of unclassified information that dwarfs the information released by the CIA.

28. *Proliferation: Threats and Response*, Department of Defense, November 1997.

29. Michael S. Serrill, "Subway Scare: Terror Takes Aim at New York," *Time*, August 11, 1997, p. 36.

30. Based in part on author's discussions with intelligence experts at the Department of State and the Japanese Embassy during August 1997.

31. "Terrorism Research Center Brief," Science Applications International Corporation, available on the Internet in July 1997.

32. See, for example, Robert David Steele, "Virtual Intelligence," Unpublished Manuscript of Open Source Solutions, Inc., Oakton, VA, May 17, 1997.

33. The official responsible for Internet hookups within CIA's intelligence division reported that due to biases against unclassified information, many all-source CIA analysts do not want Internet access, Department of State briefing, November 1997.

34. Walter Pincus, "CIA Veteran Tapped To Run Operations," *The Washington Post*, July 20, 1997, p. A16.

35. Tom Will, Defense Intelligence Agency, Open Sources Coordinator, on C-SPAN Broadcast of Conference on Government Intelligence Operations, Arlington, Virginia, September 7, 1997.

36. In the spring of 1997, the Congress called for establishment of a commission to review the activities of the intelligence community in tracking the spread of weapons of mass destruction. While this is an interesting development, it is not a substitute for the advisory committees being recommended.

37. Perl.

38. Ibid.

39. Comments by Assistant Secretary of Defense for International Security Affairs Franklin D. Kramer during a presentation to the World Affairs Council of Washington, D.C., in Arlington, Virginia, May 27, 1997. During debates in Washington on NATO expansion into Eastern Europe, one particularly articulate opponent who questioned the new rationale for NATO was Michael Mandelbaum. His report "NATO Expansion: A Bridge to the Nineteenth Century," Center for Political and Strategic Studies, Washington, May 1997, summarized his views.

40. Admiral Joseph Lopez, "NATO Is Ready," *The Washington Post*, August 12, 1997, p. A18.

41. Discussions with NATO staff members during July 1997. NATO has been supporting a small program of workshops directed to scientific cooperation related to biological weapons, but this activity is considered peripheral to the mainstream of NATO's interests.

42. *Soviet Nuclear Threat Reduction Act of 1991.*

EPILOGUE

1. "Remarks by the President on American Security in a Changing World," George Washington University, August 5, 1996, Office of the Press Secretary, The White House.

2. See also, Raphael F. Perl, "Terrorism, the Future, and U.S. Foreign Policy," CRS Issue Brief, Congressional Research Service, IB95112, November 24, 1997, p. 4.

3. Roger Medd and Frank Goldstein, "International Terrorism on the Eve of a New Millennium," *Studies in Conflict and Terrorism*, vol. 20, no. 3, July–September 1997, p. 299.

4. See, for example, Andrew Hawken, "World Must Act to Stop Genetic Weapons," MSN News report on meeting of British Medical Association, August 7, 1995. Also, in April 1997, at the request of the Geneva Foundation to Protect Health in War, professors from the University of Geneva were preparing a report on the feasibility of ethnic weapons.

5. Secretary of Defense William H. Cohen, "Keynote Address," Conference on Terrorism, Weapons of Mass Destruction, and U.S. Strategy, University of Georgia, April 28, 1997.

6. John Arquilla, "The Great Cyber War of 2002," *Wired*, February, 1998, pp. 47–62.

7. Rosemary J. Erickson, Ph.D., *Armed Robbers and Their Crimes*, Athena Research Corporation, Seatttle, Washington, 1996. This book, based on interviews with imprisoned criminals, underscores the importance of escape routes as the most important aspect in selection of crime targets.

8. Medd and Goldstein, pp. 281–316. Also, Joseph Siniscalchi, "Non-Lethal Technologies: Implications for Military Strategy," Occasional Paper No. 3, Center for Strategy and Technology, Air War College, Maxwell Air Force Base, March 1998.

9. At a meeting of the National Research Council, in Washington, D.C., in August 1997, Raphael Perl of the Congressional Research Service set forth the basis for this list of Congressional priorities.

10. Comments by Yonah Alexander at seminars on terrorism sponsored by George Washington University, February 20 and 25, 1998. These incidents were also reported in the press.

11. Ibid, see also, David Hoffman, "Russians Expanding Role in Iranian Power Plant," *The Washington Post*, February 22, 1998, p. A30.

12. "Germ Warfare," on ABC *Prime Time* featuring Diane Sawyer. February 25, 1998.

13. See, for example, Melinda Liu and Alan Zarembo, "Saddam's Secret World," *Newsweek*, March 2, 1998, p. 38.

14. For a discussion of nuclear terrorism aimed at Israel, see Louis Rene Beres, "Israel, the 'Peace Process' and Nuclear Terrorism, Recognizing the Linkages," *Studies in Conflict and Terrorism*, vol. 21, no. 1, 1998, p. 59. For comments on establishing a chemical and biological weapons free zone, see Rachelle Marshall, "Bombing Iraq Is Futile," *USA Today*, February 2, 1998, p. 12A.

Illustration Credits

1. *AS&E Bodysearch*™, American Science and Engineering, Inc. (AS&E). 1995. Billerica, MA.
2. From *Studies in Conflict and Terrorism*, 20 (3): 16, July–September 1997, Taylor and Frances Publishers, Bristol, PA.
3. Carlos the Jackal, Courtesy of Tony Stone Images© (left); Shoko Asahara and Richard Pryce, Courtesy of AP/Wide World Photos (middle and right, respectively).
4. Courtesy of the U.S. Department of Energy.
5. From *Global Proliferation of Weapons of Mass Destruction*, Part I, Senate Hearings 104–442, October 31 and November 1, 1995, page 90. Artist rendition by Ellen McLaughlin.
6. Courtesy of the New Jersey Institute of Technology, Center for Environmental Engineering and Science, University Heights, Newark, NJ.
7. Courtesy of U.S. Army, Medical Research Institute for Infectious Diseases.
8. Photographs by Glenn Schweitzer, from *Moscow DMZ*, M. E. Sharpe, Armonk, NY, 1996.
9. Courtesy of U.S. Agency for International Development.
10. Courtesy of the U.S. Customs Service.
11. Courtesy of Brian Hendler/AP Wide World Photos.
12. Courtesy of Gino Domenico/AP Wide World Photos.
13. Courtesy of U.S. Drug Enforcement Administration.

Index